Praise for Meatonomics

"Consumers can only make wise purchases of meat if the price they pay reflects the full cost of producing it—when there are no 'hidden' costs like subsidies or environmental damage. Simon is the first author to attempt a complete accounting of all these hidden costs, something that should be applauded by the vegan and meat-lover alike."

—F. BAILEY NORWOOD, PhD,
author of *Compassion by the Pound*,
associate professor, Department of Agricultural Economics,
Oklahoma State University

"This important book joins the ranks of T. Colin Campbell's *Whole* and *The China Study* in its power to expose the truth and begin to repair the health care crisis."

—PATTI BREITMAN,
co-author of *How to Eat Like a Vegetarian,
Even If You Never Want To Be One* and *How to Say No Without
Feeling Guilty*

"*Meatonomics* will grab you and not let you go. It's a critically important and absolutely fascinating and astonishing in-depth look into the devastating effects of an industry's economic take-over of our culture and our well-being. David Robinson Simon not only cogently and systematically exposes the many facets of cost externalization by the meat, dairy, egg, and fishing industries, but he also makes a compelling case for practical solutions that we can all work for, discuss, and implement, including a meat tax, changes in government subsidy programs, and personal food choices. Meatonomics has my highest recommendation—a book that liberates as it illuminates."

—WILL TUTTLE, PhD,
author of *The World Peace Diet*

"We like to think we live in a democracy, where public officials tend the general welfare. But increasingly, corporate lobbyists write our laws, and corporate interests dictate what we are allowed to know. David Robinson Simon's book is spectacularly important, because it lifts the veil and shows how the meat and dairy industries rig the game, and thus are able to stuff us with foods that imperil our health, devastate the environment, and cause unrelenting cruelty to billions of animals. He reveals the massive subsidies that make industrial meat and dairy products seem cheap, when in fact they are destroying our lives and our future. He lets us see what these industries don't want us to see— the true cost we are paying for their products. And he shows us the steps we need to take, as individuals and as a society, to restore both our economic sanity and our health."

—JOHN ROBBINS,
author of *The Food Revolution, No Happy Cows,
Diet For a New America*, and other bestsellers

"The need to transform the unhealthy, unsustainable, and unjust food system that prevails today runs deep. It will require food activists and researchers to undertake what will constitute a long march through the entire food chain. A critical starting point involves the corporate-dominated meat production system. David Robinson Simon takes us on that journey and helps us identify what we will need to confront and the changes that will need to be made."

—ROBERT GOTTLEIB,
co-author of *Food Justice*, Professor, Urban &
Environmental Policy Institute, Occidental College

"A lively, well-researched look at society's many misconceptions about the production and consumption of meat. If you eat meat, you owe it to your body and your planet to read this book."

—RORY FREEDMAN,
author of *Beg* and co-author of the Skinny Bitch series of books

"The knowledge in *Meatonomics* will free you and put you in control of your own food choices and health."

—JANICE STANGER, PhD,
author of *The Perfect Formula Diet*

CONTENTS

AUTHOR'S NOTE

Let me begin by getting a couple things off my chest. For starters, economics is subjective. John Kenneth Galbraith said the field was one in which "hope and faith coexist with great scientific pretension." The figures I propose for the costs of meatonomics are based on data that are slippery and hard to find, and the calculations themselves can vary based on how the math is done. Still, I think it's worthwhile to try. I've sought to present figures that I believe are reasonable and as accurate as possible, and in each case, to explain where they came from. Nevertheless, I'm the first to admit that this book's cost figures are, like almost everything in economics, subjective estimates.

Furthermore, while parts of this book deal with economics, medicine, and ecology, I'm not an economist, a doctor, or an ecologist. I'm a lawyer, and that's why I like to write disclaimers. A number of specialists in these areas have read and commented on the manuscript, which I hope means it contains no glaring errors. The book's analysis and conclusions are supported by research cited in more than seven hundred endnotes. Most of this information comes directly from government reports or published, peer-reviewed studies.

If you want to understand what's going on in the animal food industry, sometimes it helps to be an expert. But more often, you just need to keep your eyes and ears open, and approach the subject with what some Zen practitioners call "beginner's mind." As Zen master Shunryu Suzuki observed, "In the beginner's mind there are many possibilities, but in the expert's mind there are few."[1]

INTRODUCTION

Imagine a bakery that sells every cake, pie, or loaf of bread for a dollar less than it costs to make. It's a challenging business model, to say the least. But instead of going out of business, say the shop flourishes and expands, adding more ovens and increasing output for years. Impossible, right?

For a bakery, maybe. But not for America's big producers of meat, fish, eggs, and dairy. The animal food industry actually uses this contrarian business model with surprising success. Take hog farmers, who routinely spend an average of eight dollars more raising each pig than the animal yields when sold.[1] The farmers, at least the big corporate operators, are in hog heaven. That's because government subsidies actually make this business model profitable for those at the top. For the same reason, corporate beef producers routinely spend from $20 to $90 more than each animal's value to raise cattle.[2]

Each year, American taxpayers dish out $38 billion to subsidize meat, fish, eggs, and dairy.[3] To put this corporate welfare package in perspective, it's nearly half the total unemployment benefits paid by all fifty US states to unemployed workers in 2012.[4] However, as we'll see, unlike unemployment payments, subsidies don't actually benefit many Americans—nor many farmers—and they are often disbursed in illogical and unfair ways. Consider this: media mogul Ted Turner and former NBA star Scottie Pippen were among the more than one thousand non-farming New York City residents to pick up farming checks from the federal government in 2007.[5]

When it comes to the market for crops used as animal feed, which means the majority of crops grown in this country, America's enormous farm subsidy program turns the system topsy-turvy. Bizarrely, government handouts encourage farmers to grow more of these crops even as prices decline. This is as backward as parents giving their kids extra money to make cold lemonade in the middle of winter. It just doesn't make sense. Perhaps even worse than wasting the money, the consistent result of such a subsidy policy is to put small farmers out of business and damage rural communities here and abroad. But it doesn't end there. Taxpayers also provide subsidies to encourage fishing even when it would otherwise be unprofitable. Yet with twice the number of fishing ships patrolling the seas than are necessary for the task, humans have already destroyed one-third of the ocean's fisheries and, unless we cut back, are headed for complete destruction of all currently fished species within several decades.[6]

Few Americans are aware of the realities of meatonomics—the economic system that supports our nation's supply of animal foods— yet the peculiar economic forces powering our food system influence us in ways few imagine and nudge us to behave in ways we normally wouldn't. Among its various effects, one of the most unsettling is that the system encourages us to eat much more meat and dairy than the United States Department of Agriculture (USDA) advises.

According to conventional wisdom, factors like taste, dietary beliefs, and cultural traditions drive our decisions to buy animal foods. But the reality is that price plays a huge role in our eating choices as well. The alarming result of consumers watching our pocketbooks so carefully is that producers, who work hard to keep prices artificially low, are heavily responsible for driving demand. Doubling down on their strategy, producers also bombard shoppers with misleading messages about the need to chow down on animal foods. Consequently, Americans have, to a great extent, become puppets of the animal food industry. We eat what and how much we're told to, and we exercise little informed, independent judgment. You might think you know why you choose to eat certain foods, but as we'll see, the real reasons are much more complicated.

Spend a few hours with this book, and you'll gain vital insight into how the economics of animal food production influence your spending, eating, health, and longevity. You'll also discover how the forces of meatonomics affect the well-being of the planet and its inhabitants, including tens of billions of animals used for food, and millions of small farmers here and abroad. Learning how these forces work can help you improve your personal life and the world in so many important ways, including saving money, losing weight, boosting your health, living longer, protecting animals and the planet from abuse, and preserving rural communities in the United States and elsewhere.

Meet the Owners

The Occupy Movement knows them as the One Percent. Comedian George Carlin called them the country's Owners. They're the rich power brokers behind the scenes, the business aristocrats who own almost everything in the United States and either influence or make almost all the important decisions in the country. In the meatonomic system, the Owners enjoy a base of economic and political power practically unequaled in any other industry.

The animal food sector wields its considerable economic clout to exert enormous influence over lawmaking at both the state and federal levels. In the past several decades, animal food producers have convinced lawmakers to adopt a broad range of legislation—including some so over the top that it can only be called shocking—to protect the industry and ensure its profitability. For example, it's illegal to "defame" animal foods in thirteen states, and as Oprah Winfrey learned firsthand from a tangle with Texas beef producers, the industry does not hesitate to sue those who say unkind things about its products. Further, because undercover investigations at factory farms invariably yield graphic images of unsafe and inhumane conditions, the industry has sought—with surprising success in a number of states—to stop the flow of these shocking images by criminalizing the exposés.

Then there's the federal food bureaucracy. Meat and dairy producers have conquered the two main US agencies that oversee them—the USDA and the Food and Drug Administration (FDA)—through a process economists call "regulatory capture." This influence makes the USDA so bipolar, it's a befuddling exercise to figure out the agency's message or mission. The thirteen-member committee that formulated the agency's latest set of nutrition recommendations was tasked with looking out for the nation's health. But the group included nine members with ties to the food industry, casting doubt on the committee's good faith and on the reliability of its output.[7] In one example typical of the agency's institutional confusion, a USDA brochure advises Americans to eat less cheese, while the agency simultaneously supports advertising that urges us to eat more cheese.[8]

As for the FDA, it regularly ignores scientific research and public opinion to side with industry. In a move that might have made Louis Pasteur queasy, the agency permits milk producers to dose cows with a dangerous growth hormone (a practice outlawed in Europe and sharply criticized by a US federal appellate court). It also refuses to require labeling of genetically engineered foods despite public demand for such disclosure.[9] As the FDA moves closer to approving the sale of a new genetically modified salmon, this nondisclosure policy could soon make it impossible for consumers to distinguish between a gene-spliced fish and the real thing.

Is It Sustainable?

The animal agriculture system drives production at levels that make this sector, according to recent research by two World Bank scientists, the single greatest human cause of climate change on the planet.[10] That's right, forget carbon-belching buses or power plants; animal food production now surpasses both the transportation industry and electricity generation as the greatest source of greenhouse gases. Even worse, the system fosters financial incentives that encourage the relentless destruction of land and the routine contamination of air and water. For example, antibiotics and steroids are commonly used to make farm animals grow faster—thus yielding greater profits. (Athletes, it turns out, have nothing on cattle when it comes to artificially bulking up.) The widespread use of animal drugs means these chemicals show up not only in most of the animal foods that Americans eat but also in a majority of US waterways.[11]

Commentators have proposed a number of alternatives to improve the sustainability of animal food production. Unfortunately, these solutions generally fall short. For example, ecological rotation farming operations, like the well-known Polyface Farm (popularized in Michael Pollan's bestseller *The Omnivore's Dilemma*), represent one interesting approach to animal agriculture. However, a closer look at such farms shows a disappointing truth: they're both unsustainable and incapable of serving the demand of a nation like ours. Just addressing the local meat-eating supply of Southern California, where I live, would require thirty-three thousand farms the size of Polyface, a physical and logistical impossibility.[12]

As much as we might like our Dairy Queen and Burger King, the reality is, compared to plant protein, raising animal protein takes up to one hundred times more water, eleven times more fossil fuels, and five times more land. Without dramatic reform, the end game in the conflict between fixed resources and ever-increasing demand is likely to have a group of clear losers—the planet's inhabitants. According to Will Tuttle, author of *The World Peace Diet*, "until we are willing and able to make the connections between what we are eating and what was required to get it on our plate, and how it affects us to buy, serve, and eat it, we will be unable to make the connections that will allow us to live wisely and harmoniously on this earth."[13] Meatonomics only ratchets up the damage by artificially inflating demand and disrupting other market forces. No matter your political stripe, this should bother you. If you believe in free markets, this radical and destructive government interventionism is upsetting. If you prefer regulation, the fact that government hands your tax dollars to large corporate interests is likely aggravating. In meatonomics, there's something to annoy almost everyone.

But if we are to eat, your inner carnivore may ask, don't we *need* this food production system—despite all its quirks? Sure, we have to eat, but not like this. Americans are rational, thoughtful consumers, and we want to behave in a rational manner. But the evidence shows that artificially low prices and aggressive government messaging encourage us to consume animal foods in unnaturally high quantities. As a nation, Americans consume more meat per person than anywhere else on the planet.[14] Once, we might have celebrated our extreme consumption as evidence of good living. After all, when you hear the phrase *eat, drink, and be merry,* most people can't help but picture a few slabs of meat on the table. But today, it's one of the main reasons we have twice the obesity rate, twice the diabetes rate, and nearly three times the cancer rate as the rest of the world.[15] American longevity, once among the world's highest, now ranks fiftieth. Simply put, our heavy consumption of foods high in saturated fat, cholesterol, and other substances linked primarily or uniquely to animal foods has helped make us one of the sickest developed nations on Earth.[16]

The Price We Pay

More than any other microeconomic system in the United States, meatonomics aggressively shifts the costs of producing its goods onto American taxpayers and consumers. The only word for these costs is *staggering.* The total expenses imposed on society—that is, production costs *not* paid by animal food producers—are at least $414 billion.[17] These costs are not reflected in the prices Americans pay at the cash register. Rather, they are exacted in other ways, like higher taxes and health insurance premiums, and decreases in the value of homes and natural resources touched by factory farms.

For every dollar in retail sales of meat, fish, eggs, or dairy, the animal food industry imposes $1.70 of external costs on society. If these external numbers were added to the grocery-store prices of animal foods, they would nearly triple the cost of these items. A gallon of milk would jump from $3.50 to $9, and a store-bought, two-pound package of pork ribs would run $32 instead of $12.[18]

The American animal food industry is not alone. Most other industries distribute their profits to a relatively small group of stakeholders, and corporations commonly externalize costs in the course of generating those profits. But this industry *is* unique in the unparalleled scope of its destructive swath, the massive costs it imposes on society, and the total quantum of misery it dumps on consumers, taxpayers, workers, farmers, and animals. Consider the favorite pariah industry of many: US tobacco. Over five decades, tobacco companies were shown to have caused—and ultimately were forced to pay—$400 billion in health care costs. By comparison, as we'll see, the US animal food industry generates more than $600 billion in health care costs *every two years* and pays virtually none of them.[19] Further, unlike animal agriculture, the tobacco industry causes little ecological harm, and it's taxed—not subsidized.

Or take another sector we love to hate: Big Oil. Although the oil industry's environmental impact might rival that of animal agriculture, most petroleum products are heavily taxed—unlike animal products. Further, the $10 billion in yearly federal subsidies (including tax breaks) enjoyed by the oil industry is mere pocket change compared to the $38 billion heaped each year on the animal food industry. In the race to the absolute bottom, animal agriculture wins, hands down, as the US industry that imposes the highest economic costs on society across the board.

How Did We Get Here?

For many, this book may come as a surprise. Most of our beliefs about nutritional needs, consumption levels, and farming and lawmaking practices are based on traditions that have largely melted away—at a pace of change so slow and seductive, we're barely aware of it. As the comic strip's Calvin put it, "Day by day, nothing seems to change. But pretty soon, everything's different." Consider a few ways that the changing landscape of animal food production has both shaped the growth, and heralded the rise, of meatonomics.

For starters, forget about that bucolic *American Gothic* picture of the gentleman farmer. Industrial farming operations have largely replaced small farms, and the "pasture spring" and "little calf . . . standing by the mother" that Robert Frost saw on his family farm a century ago are lost artifacts—relics of an obsolete way of life. In the decades since 1950, American farming has undergone a major transformation, and mom-and-pop farms are mostly gone—either acquired by large corporate operations or plowed under for new housing subdivisions. For instance, between 1954 and 2007, even as demand for dairy increased by 40 percent, the number of US dairy farms plummeted from 2.9 million to 65,000.[20] We wouldn't know it from the peaceful, pastoral logos of the dairies and meat packers whose products we consume (who doesn't love a smiling cow on a package?), but today, 99 percent of the farm animals raised in the United States live in steel and concrete factories with no resemblance to a traditional farm.[21]

Then there's the fact that meat and dairy keep getting cheaper. This development is driven partly by subsidies, partly by efficient methods of factory farming, and partly by the industry's practice of offloading its costs onto others. But the upshot is the inflation-adjusted retail prices of animal foods have dropped steadily in the past century. Since 1913, in inflation-adjusted dollars, eggs have gotten cheaper by 79 percent, butter by 57 percent, and bacon by 23 percent. Here's a jaw-dropping stat: the portion of our incomes that Americans spent on meat was 2.4 percent in 1990, yet despite higher consumption levels, only 1.7 percent in 2010.[22] And of course, it's a basic rule of economics that declines in price lead to increases in demand.

Thus, the last century has also seen a significant increase in animal food consumption and its ugly cousin, obesity. Annual per-capita meat consumption has nearly doubled in the United States over the last century to its current level of 200 pounds per person.[23] Our meat and egg consumption levels are well above USDA recommendations, and this is one reason we're growing dangerously heavier. Two in three Americans are overweight and one in three is obese.[24]

But it wasn't always like this. Fifty years ago, only one in eight Americans was obese.[25] The national obesity figure increased by an average of about one-half percentage point per year for the past five decades, moving almost in lockstep with the rise of factory farming and the decline of animal foods' retail prices. Of course, higher consumption of meat and dairy is not the only reason for our nation's health issues—we also eat more sugary and processed foods than we used to—but as we'll see, volumes of research show that animal foods are a major contributing factor.

Finally, the steady rise of meatonomics has followed a disturbing, yet rampant political change: corporate influence over lawmaking has risen dramatically in the last half century. Driven largely by the expense of television advertising, the cost to get elected to US office has increased tenfold (in inflation-adjusted dollars) in the last fifty years.[26] This skyrocketing price tag has in turn dramatically boosted the amounts spent to influence lawmakers and the number of lobbyists peddling influence. (For a graphic example of how lobbying works at this level, check out the 2005 Golden Globe–nominated film *Thank You for Smoking*.) In the past three decades, as annual spending to influence Congress rose from $100 million to more than $3.5 billion (in inflation-adjusted dollars), lobbyists grew their ranks tenfold.[27]

The animal food industry is just one of many special interests to capitalize on this massive change in spending and influence, but its efforts have been particularly successful. In the past few decades, the industry has convinced lawmakers to pass scores of state and federal laws that protect animal food production in a variety of ways. These include such disturbing examples as the emasculation of dozens of laws that once prohibited cruelty to farm animals and the passage of new prohibitions against food defamation, undercover investigations, food injury lawsuits, and phantom ecoterrorism.

The Purpose of This Book

For almost as long as they've been in use, factory farms have been synonymous with three kinds of problems: environmental, nutritional, and ethical. This book proposes a fourth category: economic. We'll see how factory farming offloads massive costs onto society and how its contrarian economics drive other problems like overconsumption. Low prices are certainly not the only reason people overindulge in animal products, nor can we blame economics exclusively for the many problems associated with animal food production and consumption. Clearly, a complex set of personal and social factors are at play in our food choices and in the consequences those choices have for us and the world around us. Psychology professor Melanie Joy has proposed the term *carnism* for the belief system that drives meat consumption. This entrenched system, Joy says, "is supported by every single institution in society, from medicine to education."[28] However, while this belief system is likely responsible for persuading Americans to consume animal foods in the first place, it is in large measure the *price* of these goods that determines *how much* meat and dairy people buy. Thus, I seek to show that economic forces play a much greater role in our consumption choices than we've previously thought.

I also argue that while American consumption of animal foods is often perceived as demand driven—or spurred by consumer preferences and disposable income—it is actually heavily supply driven, or propelled by producer behavior. For instance, popular explanations for consumers' rising consumption of animal foods look at demand drivers like rising incomes and lifestyle changes.[29] But it's not that simple. The latest research shows that changes in production methods, such as the shift from traditional farming methods to low-cost industrial practices—and the resulting declines in retail prices—deserve most of the credit for the increase in consumption.[30] In other words, it is mainly producers, not consumers, who have spurred the massive increase in animal food consumption over the past century.

Moreover, state and federal governments provide key assistance in this demand-boosting process by laying out subsidies and protectionist policies that let producers sidestep the vast majority of their own production costs. Consumers get it from every side—the USDA tells us to eat more, industry tries to convince us substances like saturated fat are good for us, and lawmakers impose liability on those who might investigate, criticize, or sue meat or dairy producers. As we'll see, collectively, these meatonomic forces routinely impair the ability of consumers to make healthy decisions about what and how much to eat. These forces also cause systematic failure in the American market for animal foods.

Market failure is econo-speak for a market's inefficient allocation of goods and services which, if fixed, would yield better outcomes for all. In the following chapters, I present a three-part argument that illuminates in detail—and shows how to fix—the significant market failure caused by the economics of animal food production. First, I show that the federal government is at fault for fostering economic conditions that benefit no one except the animal food industry. With bureaucrats often turning a blind eye to how or what can be communicated to consumers, the industry engages in a sophisticated messaging campaign that is often misleading or confusing—and sometimes simply false. Regulators routinely strike out when it comes to exercising control over this and other industry activity, and through misguided legislation and policymaking, lawmakers actually encourage the industrial food complex to impose its production costs on us. When this kind of governmental negligence leads to market failure, as it does with meatonomics, the phenomenon becomes government failure.

Second, I argue that because of this government failure, the microeconomic system that produces meat and dairy is characterized by heavy overconsumption, huge inefficiencies, and massive hidden costs. This broken system damages Americans' health, hurts the environment, treats animals cruelly, and causes other harms. Moreover, these problems generate significant, measurable, financial consequences. As former US Senate Minority Leader Everett Dirksen (R-IL) famously remarked, "A billion here, a billion there, pretty soon you're talking real money." $314 billion in health-care costs. $38 billion in subsidies.* $37 billion in environmental costs. $21 billion in cruelty costs. $4 billion in fishing-related costs. Collectively, these costs would almost triple the retail prices of animal foods if they weren't offloaded instead onto consumers and taxpayers.

But the picture is not all doom and gloom. The book concludes with several suggestions to fix this broken market, restore our health, and heal the environment. On an individual level, we can each help by changing how we consume. On an institutional level, relatively simple policy changes can stimulate the economy, save 172,000 lives, eliminate $184 billion in external costs, incentivize Americans to make healthy eating choices, and cut carbon-equivalent emissions to a level not seen since 1950. This solution is practical and realistic, and because it's coupled with an income tax credit for all Americans, it's politically feasible.

---

* While a subsidy is not technically a hidden, or externalized, cost, farm subsidies are included for measurement purposes because, like externalities, they impose costs on—but provide little actual benefit to—taxpayers.

"Most of our assumptions have outlived their uselessness," said the Canadian philosopher Marshall McLuhan. We once thought the Earth was both flat and the center of the universe. A few centuries ago, we thought it wise to add lead to wine. As recently as 1929, we believed a little cocaine in our Coca-Cola was good for us.

This book asks us to challenge our assumptions about the production and consumption of animal foods—including their health effects, ethical issues, and economic impacts—and our government's role in the process. Do low prices always benefit consumers? Can we always trust peer-reviewed research? Do lawmakers and regulators really act in our best interest? Are factory farmers truly concerned for their animals' welfare? Do we actually know *why* we consume the foods we do at the levels we do? "Your assumptions are your windows on the world," says Alan Alda. "Scrub them off every once in a while, or the light won't come in."

I

INFLUENCING THE CONSUMER

1

The Brave New World of Government Marketing

In his 1932 novel *Brave New World,* Aldous Huxley imagined a future in which humans exist solely to support the economy and are conditioned from birth to buy things. Government bureaucrats manipulate the sheep-like citizens with drugs and slogans to make them consume as much as possible. In Huxley's vision, 26th-century consumers learn that "ending is better than mending" and "the more stitches, the less riches"—that is, buying new things is better than fixing old ones. But for US consumers, this eerie futuristic fantasy—with government using marketing slogans and other undue influence to drive consumption—has arrived a few centuries early. This chapter explores government marketing as a feature of meatonomics and considers its consequences for consumers.

Checkoff Programs: Unseen and Unknown, But Felt Everywhere

In the Brave New World of the 21st century—where big box stores and mega markets dominate the landscape—our government uses innocuous-sounding "checkoff" programs to encourage us to buy more animal foods and other goods. The mechanism's name persists from a time when the assessments were voluntary and producers willing to opt in participated by simply checking a box. Nowadays, the programs are tax-like and mandatory, even though the benign checkoff moniker remains.

The way they work is simple: Congress slaps a small assessment (less than 1 percent of wholesale price) on certain commodities, and the collected funds are used to pay for research and marketing programs that boost the goods' sales. So when animal food producers collect $1 per head of cattle, $0.40 per $100 of pork, or $0.15 per 100 pounds of dairy, they pass those funds on to national marketing organizations. The proceeds are allocated among state and regional industry organizations throughout the country. There aren't many Boston Tea Party–like protests when it comes to making the payments—probably because most consumers don't know about checkoffs and most producers think their trade groups put the money to good use. These trade groups don't equivocate much about what they do or why they exist. The Kentucky Cattlemen's Association, for example, keeps it simple, saying its business purpose is "Promotion of the beef industry."[1]

Although few Americans have heard of checkoff programs, we've all heard or seen the catchy, feel-good slogans they've generated:

Beef. It's What's for Dinner.

Milk. It Does a Body Good.

Pork. The Other White Meat.

Written, spoken, or sung—and flashed across every medium, including print, radio, TV, and the Internet—these statements have bombarded American consumers for decades. The echo of one particularly snappy jingle that went with a ubiquitous 1990s commercial—"The Incredible, Edible Egg"—still rattles in my brain. And while that phrase and many others predate social networking, they persist because their sticky messaging fits in perfectly with today's meme-saturated, web-dominated world. Like an ink stamp, these messages imprint themselves with authority on our subconscious and become part of our belief system. What's for dinner? Without even knowing why, many think, *Beef*.

Across the board, animal food checkoff programs are remarkably effective at making us buy more than we would otherwise. According to the USDA, for each dollar of checkoff funds spent promoting animal foods, "the return on investment can range as high as $18."[2] The beef checkoff program raises sales by $5 per checkoff dollar spent.[3] The pork checkoff program drives $14 in sales per dollar spent.[4] While it may not boast a memorable motto, the lamb checkoff provides an unusually huge boost, driving additional sales of $38, or seven extra pounds of lamb, for each dollar spent on promotion.[5] But the biggest winner might be the dairy industry, which recently boasted that over a year and a half, checkoff efforts contributed to more than 7 billion *additional* pounds of milk sold.[6] That's an *extra* forty-seven servings of dairy per person in the United States—above and beyond the hundreds of servings we would have consumed anyway during the period. Clearly, milk is up to more than just doing a body good.

All told, these programs provide funding of $557 million yearly for animal food producers to promote their goods.[7] This massive, government-mandated marketing budget gives the meatonomic system something few other microeconomic systems have: an exceedingly deep marketing war chest, deployed to boost sales of *all* goods from *all* producers in the program. A few other commodities, like cotton and soybeans, have checkoff programs of their own. Yet in every other industry, except for those lucky enough to have a checkoff program, individual corporations must fork out their own funds to increase sales rather than rely on government programs to prop up their numbers. With meatonomics, on the other hand, the effect of checkoff programs is that we all buy more of nearly every conceivable animal food than we would otherwise. Like a diner with an insatiable appetite, the animal food industry relishes the higher sales that result. Dairy promoters brag that since their checkoff program started in 1983, annual per capita consumption of milk "has climbed 12 percent to 620 pounds."[8]

Some say checkoff programs have been unfairly linked to government and are actually just the tools of good old-fashioned capitalism. They argue these checkoff arrangements involve only private firms who pool advertising monies without government participation, and their mission and methods are no different from those of any private advertiser. However, the US Supreme Court decisively rejected this position in a 2005 case involving the beef checkoff. In *Johanns v. Livestock Marketing Association,* beef industry participants who disagreed with the message of the latest beef campaign claimed that being forced to fund it violated their right of free speech.[9] The Supreme Court disagreed, holding the message was actually *government speech* (a form of speech the government can make others support). The court said:

> The message set out in the beef promotions is from beginning to end the message established by the Federal Government. . . . Congress and the Secretary [of the USDA] have set out the overarching message and some of its elements, and they have left the development of the remaining details to an entity whose members are answerable to the Secretary (and in some cases appointed by him as well).
>
> Moreover, the record demonstrates that the Secretary exercises final approval authority over every word used in every promotional campaign. All proposed promotional messages are reviewed by [USDA] Department officials both for substance and for wording, and some proposals are rejected or rewritten by the Department. . . . Nor is the Secretary's role limited to final approval or rejection: Officials of the Department also attend and participate in the open meetings at which proposals are developed.[10]

This crystal-clear language from the highest court in the land leaves little doubt that the beef checkoff program, and the messages it generates, are the product of the federal government. Simple logic shows that other animal food checkoff programs, which were established by Congress in the same way and are similarly administered by the USDA, are equally the mouthpieces of the federal government. So when one of these organizations speaks—regardless of the product it's hawking—it may say it's the National Pork Board, but the background sounds you're hearing are the imposing bass tones of the US government.

In fact, the government's continued regulatory involvement is a necessary component for mandatory checkoffs to remain legally and operationally viable. If Congress simply created a checkoff program and then stepped aside to let industry run it, the First Amendment's free speech protections would likely prevent the industry majority from bullying dissenters into participating in its message.[11] Under those circumstances, forget the government speech exception: it wouldn't apply and individual participants could opt out. The result would be a checkoff program that is in fact optional, not mandatory.

Why does that matter? Because such a scenario would likely undercut the force of the messaging. As research on optional checkoffs shows, economic free riders—those group members who opt out of paying for all the snazzy commercials but still enjoy their benefits—significantly lower the effectiveness of such programs.[12] Ultimately, a lack of government involvement would likely lead to the decline—or maybe the end—of checkoffs.

Checking Out Checkoffs

Few people have heard of checkoffs, and fewer still have considered their effects. Yet these programs have a number of important consequences, some good and some bad, that merit attention. First and foremost, checkoffs stimulate the economy. By boosting sales, checkoffs create jobs and drive spending. As the USDA puts it, "The fundamental goal of every checkoff program is to increase commodity demand, which increases the potential long-term economic growth of all sectors of the industry and the communities in which they operate."[13]

With a few calculations, we can estimate the overall economic effect of checkoffs. It's a full-fledged bonanza: As table 1.1 shows, the USDA's figures for return on investment from checkoff funds suggest that checkoffs boost sales of animal foods by about $4.6 billion.[14] There's also a multiplier effect related to this sales increase: checkoffs create new jobs, and that in turn increases spending. Applying the typical multiplier used by researchers (0.77) to the sales total yields $8.2 billion in total economic stimulus related to animal food checkoffs.[15] Not bad, but what about the other side of the ledger?

For starters, animal food production generates large external costs—expenses that producers impose on society instead of paying themselves. In the book's second half, we'll see that for each $1 of animal food sold at retail, the industry generates about $1.70 in external costs. Applying this ratio to the $4.6 billion sales figure reveals that checkoffs generate roughly $7.8 billion in external costs *not reflected* in the retail prices of the goods they promote. That's nearly equal to the economic activity they generate. As with many of the interesting equations that meatonomics presents, the $64,000 question is whether the trade-off is worth it.

TABLE 1.1 Effects of Checkoff Spending on Animal Food Sales (dollar amounts in millions)[16]

ANIMAL FOOD CHECKOFF PROGRAM	ANNUAL CHECKOFF FUNDS SPENT	RETURN ON CHECKOFF FUNDS INVESTED	EXTRA SALES FROM CHECKOFF SPENDING
Pork	$65.4	14	$915.6
Beef	79.8	5	399.0
Eggs	21.0	6	126.0
Lamb	2.3	38	87.4
Milk	107.8	8	862.4
Dairy	281.2	8	2,249.6
Total	$557.5		$4,640.0

Checkoffs, moreover, cause us to buy more animal foods than we would otherwise. Yet judging from the data, Americans already eat plenty of these foods and don't need more. Teenagers, for example, consume 78 percent more saturated fat and 48 percent more cholesterol—both linked primarily or exclusively to animal foods—than government guidelines recommend.[17] One in three US teenagers is obese or overweight, triple the rate in 1963, and a growing number have diabetes or high blood pressure—diseases directly linked to meat and dairy consumption and formerly seen only rarely before adulthood.[18]

Nevertheless, the USDA keeps urging these kids to eat more of the very foods that help make them fat and unhealthy. The huge milk promotion Fuel Up to Play 60, for instance, enjoys more than $50 million yearly in government-mandated funding and reaches 36 million students in seventy thousand schools.[19] And checkoff funding helped the Dairy Board team with Domino's Pizza to offer pizzas in two thousand US schools.[20] Yet it's not just kids who overindulge; as table 2.1 in chapter 2 shows, adult Americans also routinely consume more animal foods than the USDA recommends.

Weird Science

With annual promotional funds of $389 million, the dairy industry enjoys nearly three times the checkoff spending of all fruit and vegetable producers combined (not to mention a marketing budget that would be the envy of many a Hollywood studio).[21] To look at it another way, dairy spends more on advertising in one week than the blueberry, mango, watermelon, and mushroom industries spend together in a year.[22] Under federal law, checkoff funds are intended to be used for both promotion and research. Thus, the National Dairy Council, the largest of dairy's many checkoff-funded arms, boasts that it "partners with top universities and other research facilities across the United States to support nutrition research efforts."[23] Dairy research, funded by at least $58 million yearly, is largely focused on finding ways to convince consumers that dairy is healthy.

Since industry-funded research might be suspect, dairy takes steps to ensure its research appears unbiased. For scientific credibility, research must be published in a respected, peer-reviewed journal. But here's the rub: the National Dairy Council ensures access to such journals, and the benevolence of their editorial boards, by donating cash to a number of nutritional organizations. These include the American Society for Nutrition (whose other corporate sponsors include Dannon and McDonald's) and the Academy of Nutrition and Dietetics (brought to you by the National Cattlemen's Beef Association).[24] Both organizations publish prestigious research journals.

The "best source for the most accurate, credible and timely food and nutrition information," boasts the website of the Academy of Nutrition and Dietetics. But what's left unsaid is the Academy, formerly known as the American Dietetic Association (ADA), has a particularly cozy relationship with dairy. As the world's largest organization of food and nutrition professionals, with over seventy thousand members, it's easy to see how food industry players can benefit from access to this influential group. This begs the question, just how accurate and credible *is* the organization's nutrition advice?

In a 2007 press release discussing a major increase in the size of the National Dairy Council's funding commitment, the ADA said the sponsorship arrangement gave dairy producers "prominent access to key influencers, thought leaders and decision-makers in the food and nutrition marketplace."[25] The release went on to illustrate, with candor, how the relationship benefits the Dairy Council. One quid pro quo of past sponsorship apparently included the ADA's endorsement of the Dairy Council's "3-A-Day of Dairy" campaign, which educates consumers and health professionals about the nutrition and health benefits of consuming three servings of fat-free or low-fat milk, cheese and yogurt a day."[26] The ADA release didn't disclose the extent of the Dairy Council's generosity, but judging from the size of other contributions from animal food producers to nonprofits, it's safe to assume it wasn't insignificant. The National Livestock and Meat Board, for example, gave $189,000 in one year to the American Heart Association.[27]

Dairy also seeks to extend its scientific influence by installing its people on boards, committees, and editorial panels of nutritional organizations and their journals. One of these people is Gregory Miller, who serves in multiple capacities—president of the Dairy Research Institute, executive vice president of the Dairy Council, and committee chair for the American Society for Nutrition.

Miller and I spoke about dairy research. Among other things, I was curious about studies that have looked at industry influence in the scientific process. These studies find that industry-funded research is up to four times more likely to reach conclusions favorable to the sponsor than unfavorable.[28] In one of these studies, researchers found that "systematic bias favors products which are made by the company funding the research."[29]

According to Miller, the dairy industry provides a sort of public service through its support of nutrition research. "With government funding continuing to shrink," Miller told me, "industry has a responsibility to help fund some of the research that needs to be done out there." In light of such apparently selfless motives, who could accuse the dairy industry of bias? Furthermore, Miller assured me dairy research is *not* biased, twice using the Fox News slogan "fair and balanced" to drive home the point.

But what about the study that found industry-supported research is four times more likely to reach conclusions favorable to its sponsor? "That study design is somewhat flawed," Miller told me. "I would take it with a grain of salt."

Miller sent me a number of published articles from industry-funded research. These studies have titles like "Dairy Calcium Intake, Serum Vitamin D, and Successful Weight Loss" and, even catchier, "Drinking Flavored or Plain Milk is Positively Associated with Nutrient Intake and Is Not Associated with Adverse Effects on Weight Status in US Children and Adolescents." For anyone interested in just how fair and balanced this research is, a look at one study is enlightening.

In 2010, researcher Patty Siri-Tarino of the Children's Hospital Oakland Research Institute and three colleagues published an article that found consumption of saturated fat does *not* cause heart disease.[30] This article's surprising conclusion runs contrary to a significant and consistent line of published research that finds exactly the opposite—that dietary saturated fat causes heart disease.[31] Not surprisingly, the news that eating fat doesn't lead to heart disease hit the blogs like celebrity wedding gossip. The animal food industry now trumpets the Siri-Tarino study as one of several said to debunk the "myth" that saturated fat is unhealthy.[32]

Unfortunately for those of us who love fatty foods, this news doesn't call for a celebratory pizza. The Siri-Tarino study suffers from what many research scientists consider a defect in methodology: failure to appropriately control for an important confounding factor. Siri-Tarino's article is a meta-study—that is, it compiles and evaluates research from a number of studies to reach an empirical conclusion. The saturated fat studies analyzed in Siri-Tarino all use the cohort research model, which compares different groups to determine their incidence of disease over time. In any cohort study, confounding factors that could skew the results must be controlled. For example, because elderly people have a categorically higher incidence of heart disease than children, comparing a high-fat eating cohort of octo-genarians to a fat-free group of teenagers would be misleading—we wouldn't know if the teenagers' lower heart disease rate was related to their age or their diet.

The studies assessed by Siri-Tarino generally identified and adjusted for a number of confounding factors, such as age, gender, and lifestyle. So far, so good. But they *did not* adjust for the single most important confounding factor that any study of the health effects of fat or cholesterol in an animal food must: consumption of other animal foods. All animal foods routinely contain both saturated fat and dietary cholesterol, and surprisingly, low-fat animal foods like chicken and salmon are actually chock-full of cholesterol.[33] Accordingly, any study that seeks to assess the effect of saturated fat on health *must control for the confounding effect of dietary cholesterol.* The best way to do this is to compare a group whose members eat both saturated fat *and* cholesterol with a group whose members eat less fat and *no* cholesterol.

Because dietary cholesterol is present in animal foods but not in plant foods, a low-fat research cohort should eat only plant foods. Otherwise, low-fat eaters of animal foods can easily soak up cholesterol at levels equal to or greater than those in the high-fat cohort, and the low-fat diet will show little or no health difference when compared to the high-fat diet. Significantly, the research analyzed by the Siri-Tarino meta-study did not control for this cholesterol factor. In other words, there was no meaningful baseline for comparison since members of all cohorts consumed animal foods, and these all contained dietary cholesterol. As a result, it is little wonder the study found that low-fat diets and high-fat diets present comparable risks of heart disease. When you consider that chicken and salmon contain the same amount of cholesterol as ground beef (but less fat), it's not hard to see why Siri-Tarino's research found what it did.[34]

Many in the scientific community question such research methodology. One critic is T. Colin Campbell, professor emeritus at Cornell University, lead researcher in numerous nutritional studies, and coauthor of *The China Study.* Campbell notes the practice of replacing high-fat animal foods with low-fat animal foods, which is common in the studies analyzed by Siri-Tarino: "If one kind of animal-based food is substituted for another, then the adverse health effects of both foods, when compared to plant-based foods, are easily missed."[35] Discussing the Nurses' Health Study, a well-known study analyzed in Siri-Tarino, and which employed methodology typical of Siri-Tarino's other subject studies, Campbell writes:

> It is the premier example of how reductionism in science can create massive amounts of confusion and misinformation, even when the scientists involved are honest, well-intentioned and positioned at the top institutions in the world. Hardly any study has done more damage to the nutritional landscape than the Nurses' Health Study, and it serves as a warning for the rest of science for what not to do.[36]

It may be irrelevant that the National Dairy Council helped fund the Siri-Tarino study by paying two of the study's four authors, including lead author Siri-Tarino.[37] That's because the meta-study merely compiled results from dozens of other studies—many of which were not industry funded. Then again, even in such a meta-study, the criteria used for study selection can be highly subjective.

Further, the Siri-Tarino article appeared in *The American Journal of Clinical Nutrition*, published by the American Society for Nutrition. The journal frequently publishes research which concludes that animal foods are unhealthy, and many of those articles are cited in this book. On the other hand, as we've seen, the journal's overseeing organization is sponsored by the Dairy Council as well as others in the animal food industry and features an industry executive in one of its highest committee posts.

Regardless of whether Siri-Tarino is objective science or industry whitewashing, the importance of such research to animal food producers is clear. Research like Siri-Tarino can be highly effective at boosting product sales—in fact, sometimes even more effective than buying billboard space and running TV ads. One group found that $1 spent on pork research yielded a $25 return, while $1 spent on promotion returned only $8.[38] Perhaps Miller's claim of industry generosity is overstated. Perhaps checkoff programs in fact spend millions on research not to help the public but merely to find new ways to increase sales or reduce costs. One might well conclude that the prudent approach to such industry-funded research is to treat it as you would a carton of expired milk—with caution.

The Dubious Value of Checkoffs

Do we need checkoffs? They stimulate the economy, but in the case of animal foods, they generate almost $2 in external costs for every $1 of stimulus. Some promote indisputably healthy foods like blueberries and mangoes; others encourage those who already eat too much animal foods to eat more. Perhaps the fundamental issue surrounding checkoffs is the one Huxley raised: whether it's appropriate for government to urge its citizens to buy things they don't really need.

Take animal foods off the table for a moment. In some ways, the question of government influence is as relevant for peanuts and popcorn, both of which have checkoff programs, as it is for meat, eggs, and dairy. Is it right for the state to use cute corporate mascots like Poppy and Captain Kernel to cajole us to buy more popcorn at the movies? Recall that this is not conventional advertising, in which one private firm tries to convince us its products are better than another's. This is government-sponsored, across-the-board demand boosting—designed to sell more of everything in the category. As the enabling legislation explains, checkoffs are meant to "increase the overall demand" for the goods they cover.[39]

These are the first few notes of a motif that repeats throughout this book like a songbird's call. As H. L. Mencken quipped, "When they say it's not about the money, it's about the money." The federal government doesn't promote food to boost consumers' health. After all, many of the foods we're urged to buy are bad for us—particularly at the levels at which we're told to consume them. So why do our elected and appointed representatives tell us to eat more? Because industry demands it as a way to increase sales, and as we'll see, industry usually gets what it wants.

Not all checkoffs are created equal. Except for those covering animal foods, most checkoffs generate relatively low external costs. For checkoffs that promote low-impact goods like mangoes, blueberries, and mushrooms, perhaps the value of economic stimulus outweighs the creepiness associated with government influence. Or perhaps not. But at any rate, animal food checkoffs are in a herd of their own. They encourage Americans to consume meat, eggs, and dairy at much higher levels than normal. They drive disproportionately high external costs. The worst part, as we'll see in chapter 6, is that checkoffs help to sicken an already-ill nation. Maybe the question we should ask ourselves about these programs is: Got Milked?

Food for Thought

- Although Americans already consume much more animal foods than the USDA recommends, that agency continues to oversee checkoff programs that spend $557 million yearly urging us to buy and eat even more of these foods.

- Checkoffs are remarkably effective. By funding aggressive marketing and research programs that convince consumers that animal foods are a healthy and necessary part of our daily diet, checkoffs boost sales by as much as $38 for each checkoff dollar spent.

- Without government involvement, checkoffs would be dramatically less effective and perhaps even nonexistent. If the USDA disengaged from checkoffs that promote animal foods, it would significantly reduce the nation's routine overconsumption of these foods.

2

Massaging the Message: Shaping Consumer Beliefs

A deadly strain of swine flu raced across North America in the spring of 2009, infecting one in five Americans and hospitalizing a quarter of a million people. As sick air travelers rapidly spread the virus throughout the rest of the world, the World Health Organization issued its highest level pandemic warning. In the United States, a determined group of animal food producers and government officials sprang into action to address the crisis. But this wasn't the type of emergency-response coalition you might expect: its focus was on saving profits, not people. As a spokesman for the National Pork Producers Council warned, "This flu is being called something that it isn't, and it's hurting our entire industry. It is not a 'swine' flu, and people need to stop calling it that . . . they're ruining people's lives."[1]

For years, the animal food industry has sought—successfully, in most cases—to mold consumer attitudes. We've seen how well-funded government checkoff programs urge us to buy more meat and dairy. But checkoffs are just one piece of meatonomics' multipronged messaging machine. This chapter explores some of the other ways the industry shapes American attitudes toward meat and dairy to keep itself in the black.

Of course, product marketing is as American as apple pie. Animal food producers are certainly not the only manufacturers to use sophisticated marketing and public relations tactics to sell their goods. But there is something unique about the messaging that *this* industry disseminates. Frequently, as we'll see, communications from industrial animal farmers lack an essential element that consumers expect *and* the law requires: fair play. In some cases, their words are technically accurate but nevertheless unfair or misleading. In others, they're just plain wrong. And when people rely on incorrect or deceptive information about food, the results can be downright dangerous to their health. As with many of the industry's other characteristics explored in this book, this shady approach to salesmanship often puts meat and dairy producers in a class of their own. Honesty only makes sense, said Mark Twain, "when there's money in it." When it comes to promoting animal foods, it seems dishonesty pays.

A Flu by Any Other Name

The pork industry's fear of lives "ruin[ed]" from the 2009 flu pandemic came to pass in ways more literal than envisioned. After suffering advanced flu symptoms like chills, fever, coughs, vomiting, and diarrhea, some twelve thousand Americans ultimately died from swine influenza.[2] While a variety of state and federal agencies worked to help the sick and educate the public, the US Department of Agriculture turned its nursing skills to the pork industry's financial health. At a press conference in April 2009, USDA Secretary Tom Vilsack reassured a worried nation, "There are a lot of hardworking families whose livelihood depends on us conveying this message of safety . . . and we want to reinforce the fact that we're doing everything we possibly can to make sure that our hog industry is sound and safe. . . . This really isn't swine flu. It's H1N1 virus."[3]

Governmental, educational, and medical institutions quickly adopted the official name change. At the health care company where I work as general counsel, amid some confusion as to what to call the disease threatening our elderly patient base, our chief medical officer instructed all personnel to begin using the new name immediately. Yet the sudden and unexplained name change caught some observers off guard. "H1N1? . . . Huh?" asked one ABC News reporter, "Not swine flu? . . . What changed?"[4]

For trivia buffs interested in which name is actually correct, H1N1 or swine flu, the answer is both. H1N1 is clinically acceptable. But swine flu is also accurate, notwithstanding the pork industry's claims—later shown to be false—that no pigs were involved. A group of scientists around the world studied the origins of the flu pandemic and published their results in the June 2009 issue of *Nature*. The thirteen scientists concluded that the disease started "in swine, and that the initial transmission [from pigs] to humans occurred several months before recognition of the outbreak." These findings, according to the scientists, "highlight the need for systematic surveillance of influenza in swine, and provide evidence that the mixing of new genetic elements in swine can result in the emergence of viruses with pandemic potential in humans."[5] The official name change likely helped the pork industry avoid further losses, but it also deflected legitimate attention from the real source of the problem: sick pigs.[†]

What's in a Name?

In an age when any news—good or bad—travels at the speed of light, negative headlines can hurt an industry in real and lasting ways. For pork producers already reeling from years of losses related to high feed costs, the swine flu debacle could have been much worse. Consider a recent naming crisis that hit one producer especially hard: the pink slime episode of 2012. Known to the industry as lean, finely textured beef, pink slime is a protein paste created by spinning otherwise-inedible beef scraps in a centrifuge and treating it with ammonia to kill bacteria. US law allows the product to be added to ground beef as filler in amounts up to 15 percent without additional labeling. After ABC News reported in March 2012 that more than two-thirds of ground beef in the United States contained pink slime, public response was swift and merciless. The nation's largest grocery chains—Supervalu, Kroger, Safeway, Stop & Shop, and Food Lion—announced they would no longer sell ground beef containing the additive.

---

† In fact, influenza is just one of many infectious diseases that start in animals and spread to humans. Other infectious, zoonotic diseases include avian flu, bubonic plague, cholera, dengue fever, Ebola, herpes, HIV, leprosy, measles, SARS, smallpox, tuberculosis, West Nile virus, and yellow fever. H. Krauss et al., *Zoonoses: Infectious Diseases Transmissible from Animals to Humans* (Washington, DC: ASM Press, 2003).

Striking faster than Lindsay Lohan's PR people after one of the starlet's run-ins with the law, Beef Products Inc. (BPI), the product's manufacturer, fought back. Three state governors—Rick Perry of Texas, Sam Brownback of Kansas, and Terry Branstad of Iowa—toured a BPI plant with reporters and munched on burgers containing the controversial additive. T-shirts were printed with the odd slogan, "Dude, It's Beef." Government officials even pitched in; one Georgia bureaucrat begged media to use the product's "proper name."[6] But it was too little, too late—public opinion had formed and would not be changed. As demand for pink slime plummeted, BPI was forced to close three of its four plants.[7] If a manufacturer can be pushed to the brink this quickly, it's easy to see why meat and dairy producers pursue messaging with such determination.

It takes lobbying to convince government to change the name of a disease or to officially promote a consumer-friendly product name. However, because checkoff organizations are legally prohibited from lobbying, a variety of special-purpose organizations handle the task. More than a dozen politically focused trade groups apply pressure to our state and federal lawmakers on issues like what to call food. Because some of these groups aren't required to file tax returns, it's hard to gauge their spending. For those groups who do file returns, annual spending totals more than $138 million.[8] Added to the spending by checkoff programs, that brings the tally to nearly $700 million in total documented yearly spending by animal food trade groups to spread their message or influence lawmakers. Of course, this excludes outlays by individual producers like Cargill, Smithfield, Tyson, and others. It also excludes some large trade groups whose financial statements are not public. One of these, for example, is the United Egg Producers—the main trade group representing the $6.5 billion US egg industry.[9]

Hearts and Minds

In the battle for the souls of the nation's eaters, it's not enough merely to disseminate a message. Like a wartime radio monitor, one must also carefully analyze and decode messages from the opposition.

The industry approaches its task with vigilance. Regularly assessing consumer attitudes toward meat and dairy, industry monitors are quick to take corrective steps when needed. With that directive in mind, the National Pork Board recently surveyed kids to determine whether they had been influenced by animal advocacy organizations. Surprisingly, more than half the kids had heard of such groups, and one-fourth said that messages from the groups had influenced their eating habits. "We're keeping a close eye on these activist groups and their messages," says Traci Rodemeyer, director of pork information for the National Pork Board, "and we're prepared to take action if they escalate their efforts to target children."[10]

In 2010, the American Meat Institute (AMI), one of the main trade organizations promoting the meat industry, surveyed adult Americans to gauge their attitudes toward a set of troubling messages circulating among the meat-buying public.[11] Among others, the messages included the factually accurate propositions that eating red meat increases heart disease risk and that Americans eat more meat than recommended. For meat producers, the survey results were cause for alarm. Consumers voiced concern about these and other issues, and the meat industry got a favorability score of only 48.7 out of 100. (That's almost two points worse than the mediocre approval ratings of steroidal slugger Barry Bonds.[12]) AMI sprang into action. "A multiple media curriculum was developed to debunk these myths," wrote AMI's director of scientific affairs Betsy Booren. "The messaging was factual, positive, and consumer and media friendly."[13]

One of the consumer-friendly tools AMI deployed in its campaign is the *Meat Mythcrushers* website, which seeks to correct the "myths and misinformation" Americans learn about food from "news media, books and movies."[14] The site serves as a showcase for some of the industry's most interesting messaging tactics. For example, we learn that a "very large 2010 study" showed dietary saturated fat *does not* cause heart disease. The study, of course, is the dubious Siri-Tarino paper discussed in the prior chapter.

Even more interesting is the site's treatment of a well-documented problem: American overconsumption of meat. The site purports to debunk the overeating "myth" by pointing out that men's and women's daily consumption levels of meat and poultry, averaging 6.9 ounces and 4.4 ounces, respectively, are well within the daily range of "five to seven ounces" recommended by the USDA.[15] If you've ever wondered about the meaning of *truthiness,* the term popularized by Stephen Colbert, this assertion nails the definition. While it has certain elements of truth, the message is downright false in one respect and misleading in another. First, USDA guidelines recommend a maximum of 6.5 ounces, not seven, from the protein foods group per day.[16] It's easy to see why AMI chose to round up to seven: this allows the male consumption figure of 6.9 to squeeze just inside the range.

Worse, the AMI's daily consumption figures exclude two common animal foods listed in the USDA's protein foods group: fish and eggs. Table 2.1 shows the effect of adding these foods to the calculation. In fact, in virtually every American demographic, consumption of animal foods is well *above* USDA recommendations. Note that this table does not reflect *total* protein consumption—those figures are higher still because they include vegetable proteins like nuts and beans, and dairy, which the USDA oddly doesn't treat as protein and insists on placing in its own category for recommended daily servings. Nevertheless, the data for recommended and actual daily servings for meat is alarming. For males between twenty and fifty-nine, who routinely eat from one-third to two-thirds more than the daily recommended amount of meat, eggs, and fish, the AMI's reassuring suggestion that they're eating the right amount is downright dangerous.

TABLE 2.1: US Daily Recommended and Actual Consumption of Meat, Eggs, and Seafood[17]

SEX/AGE	RECOMMENDED (OZ)	CONSUMED (OZ)	EXCESS CONSUMED
Male			
20–29	6.5	8.8	36%
30–39	6.0	9.9	66%
40–49	6.0	9.3	56%
50–59	5.5	7.9	44%

SEX/AGE	RECOMMENDED (OZ)	CONSUMED (OZ)	EXCESS CONSUMED
Female			
20–29	5.5	6.0	9%
30–39	5.0	6.0	21%
40–49	5.0	5.9	18%
50–59	5.0	5.4	8%

Sometimes industry's aggressive messaging tactics lack even the thin veneer of truthiness. In 2011, People for the Ethical Treatment of Animals (PETA) sued the California Department of Food and Agriculture and the California Milk Advisory Board over the latter's claims that dairy cows in California are "happy." This message is central to the Milk Board's advertising campaign, which seeks to distinguish California dairy products from out-of-state goods based on the premise that cows are happier in California than elsewhere. In selling this point, the promotional messaging claims, "California dairy cows live happy all year long," and California dairy producers "work day in and day out to ensure their cows are healthy and comfortable."[18] Commercials promoting these assertions feature grassy valleys, rolling hills, and cows grazing freely in huge pastures. They end with a voiceover: "Great milk [or cheese] comes from Happy Cows. Happy Cows come from California. Make sure it's made with Real California Milk [or Cheese]." In fact, it seems that few American dairy cows have much reason to be happy (whether they live in California or elsewhere). According to research cited in the lawsuit, dairy cows raised in US factory farms (where most American milk is produced) routinely encounter a variety of difficult circumstances including:

- Johne's disease, a chronic wasting illness that affects two-thirds of all US dairy cows and causes severe weight loss and diarrhea.

- Routine branding, and the burning of budding horns—both done without anesthesia, which peer-reviewed studies and the American Veterinary Medical Association have described as "acutely painful and stressful" for the cows.[19]

- Spending most of their lives crammed into tiny, concrete-floored stalls fitted with brisket boards, which prevent the animals from reclining comfortably.[20]

- Aggressive milking quotas that make industrially raised cows more likely to "die prematurely and/or suffer from lameness, mastitis, respiratory disease, metabolic problems, reproductive complications and other sicknesses than 'normally' producing dairy cows."[21]

In fact, despite such widespread reasons for bovine unhappiness, it seems that most dairy cows would nonetheless be "happier" almost anywhere *other than* California. Specifically, California cows "have a statistically greater chance of experiencing discomfort, suffering from painful diseases and/or of dying prematurely, than cows elsewhere in this country."[22]

It just doesn't sound like any of the nation's industrially raised dairy cows, least of all those in California, have much reason to be happy. PETA tried to bring this lawsuit for more than a decade, although the first few complaints they filed were dismissed for procedural reasons. In the latest filing, PETA won an important, early-round victory when the judge ordered the defendants to release thousands of pages of documents claimed to be trade secrets. At this writing, the lawsuit is still pending.

Hitting Below the Belt

In 2008, the Physician's Committee for Responsible Medicine (PCRM) launched a campaign to educate people about the health dangers of processed meats. As a matter of fact, dozens of studies published in peer-reviewed medical journals establish that eating processed meats is linked to cancer.[23] But rather than address this copious research, AMI responded with a press release designed to deflect attention from the underlying science. The release's main theme was evident from its title: "Media Needs to Check Background of Pseudo-Medical Animal Rights Group and Cease Coverage of Alarmist and Unscientific

Attack on Meat Products."[24] Such an ad hominem, or to-the-man, personal attack is a logical fallacy that ignores the merits of the message itself. Stanley Cohen, professor of sociology and author of the book *States of Denial,* calls this a denial technique used by those who:

> try to deflect attention . . . to the motives and character of their
> critics, who are presented as hypocrites and disguised deviants.
> Thus the police are corrupt and brutal, teachers are unfair and
> discriminatory. By attacking others, the wrongfulness of [one's]
> own behavior can be more easily repressed or lost to view.[25]

AMI is no stranger to this tactic. In another example, the trade organization issued a press release in 2001 responding to a petition that various groups had filed with the USDA asking the agency to enforce the federal Humane Methods of Slaughter Act. Rather than address the petition's many eyewitness accounts of inhumane practices in the slaughter process, which were signed under penalty of perjury, AMI attacked the petitioner organizations themselves. "It is important to note," said AMI, "that the credibility of some of the petitioners is in serious doubt."[26]

AMI is not alone in seeking to deflect attention from legitimate, anti-industry messages by questioning the messengers' motives, credibility, or integrity. In the next chapter, we'll see further evidence of institutional efforts to marginalize those whose messages could harm industry. For example, animal agribusiness has persuaded Congress and most state legislatures to pass legislation that brands those who interfere with factory farms as "terrorists." This is a powerful method not only of curbing activism, but also of discrediting activists and their message.

Protein Preaching

Is animal protein a life-enhancing elixir? From a young age, we're taught it fosters health, growth, vitality, virility, and sometimes even weight loss. The alternative to getting plenty of it, we're told, could be protein deficiency. Never mind that the typical American has never had—nor ever will have—protein deficiency and has little idea what its symptoms might be. We've heard of it, we're scared of it, and whatever the heck it is, we don't want it.

Spurred by the most basic force of meatonomics—the drive to sell more meat and dairy—animal food producers use our protein fears to their advantage. For example, a beef checkoff website suggests when deciding how much meat to eat, we go beyond the bare minimum needed to "prevent protein deficiency."[27] Elsewhere on the site, we're warned:

> HEALTH ALERT: Sarcopenia.
>
> Sarcopenia is a condition associated with a loss of muscle mass and strength in older individuals. . . . While there is no single cause, insufficient protein intake may be a key contributor to this condition.[28]

The key phrase here is *may be.* In fact, the research linking sarcopenia to protein deficiency is spotty and inconclusive. A 2001 study published in *The Journal of Laboratory and Clinical Medicine* found simply, "Decreased physical activity with aging appears to be the key factor involved in producing sarcopenia."[29]

We're regularly bombarded with protein messages like these. How accurate are they? What are the health consequences of following them? Because protein is such an important nutrient, and emerging research presents an array of new findings on the subject, it's worthwhile to assess the protein messages that influence our consumption habits.

Where Do You Get Your Protein?

Here's something to chew on: a peanut butter and jelly sandwich on whole wheat bread contains more protein (14 grams) than a McDonald's hamburger (13 grams). Many consumers think plant foods contain little protein—in any case, not enough to meet our daily needs. But a closer look suggests the animal food industry may be overhyping animal protein in ways that are clinically unsupported.

For humans, the best guidance on protein requirements is contained in a 284-page report produced jointly by the United Nations and the World Health Organization (WHO).[30] According to this report, an adult needs 0.66 grams of protein per kilogram of body weight per day.‡ For a 170-pound adult, this is about 50.8 grams of protein per day. An omnivore could fill this quota with just one chicken breast and one drumstick per day, although among American consumers, such restraint is rare. Males between twenty and fifty-nine, for example, typically consume more than 100 grams of protein daily—twice the level recommended by WHO.[31]

With 50.8 grams of protein (adjusted by individual body weight) as a rough daily target, we can evaluate the meatonomics claim that it's hard to obtain adequate protein without eating animal foods. Consider these surprising protein equivalents, courtesy of the USDA: a baked potato contains as much protein as a hot dog, 2 ounces of peanuts equals a chicken pot pie, and ounce-for-ounce, roasted pumpkin seeds have more protein than ham. As table 2.2 shows, many plant foods contain protein at levels equal to the same or even larger amounts of animal foods.[32]

TABLE 2.2 Protein Equivalents in Animal and Plant Foods[33]

PROTEIN (G)	ANIMAL FOOD	PLANT FOOD
21	Double cheeseburger (w/ condiments)	Trail mix (1 cup)
18	Ham (3 oz., extra lean, canned)	Pumpkin seeds (2 oz., roasted)
16	Crab meat (3 oz., cooked)	Peas (1 cup, split, boiled)
13	Chicken pot pie	Peanuts (2 oz., roasted)
9	Turkey (1 patty, breaded, fried)	Hummus (1/2 cup)
6	Egg (large, hard-boiled)	Pistachios (1 oz., roasted)
5	Frankfurter (beef)	Potato (baked)

---

‡ The USDA also issues recommendations regarding protein consumption, although its guidance is substantially higher. The agency recommends 0.8 grams of protein per kilogram per day, which works out to 61.7 grams for a 170-pound adult or about 20 percent more than UN/WHO. I use the UN/WHO recommendations because they're more consistent with current research and less likely to be influenced by industry (*see* chapter 4).

PROTEIN (G)	ANIMAL FOOD	PLANT FOOD
4	Cheese (1 oz., feta)	Grapefruit juice (6 fl. oz., from concentrate)
2	Ice cream (1/2 cup, vanilla)	Blackberries (1 cup)
1	Cream cheese (1 tbsp.)	Cocoa (1 tbsp., dry, unsweetened)

In fact, every fruit, vegetable, nut, seed, or grain we put in our bodies has protein—in most cases, at surprising levels. You like to kick back with a Budweiser? A can of beer contains 2 grams of protein. A basic salad doesn't seem hardy enough to add a bit of muscle? A cup of romaine contains a gram of protein. In fact, calorie for calorie, green vegetables like kale, broccoli, and romaine lettuce contain twice as much protein as steak.[34] As one team of experts noted, "It is difficult to obtain a mixed vegetable diet which will produce an appreciable loss of body protein."[35]

A recent poll found that nearly 16 million Americans are vegetarian (that is, they eat no meat) and of these, nearly 8 million are vegan (that is, they eat no animal products whatsoever).[36] Yet there is no clinical evidence that members of either group suffer from protein deficiency. In fact, a number of commentators note that protein deficiency is largely associated with caloric deficiency, and for anyone consuming sufficient calories, adequate protein is not a concern.[37] In a report that is the basis for the USDA's protein recommendations, the National Academy of Sciences downplays the risk that people on a plant-based diet lack sufficient dietary protein. According to the National Academy, "available evidence does not support recommending a separate protein requirement for vegetarians."[38]

Nevertheless, the animal food industry promotes the message that plant protein is lower in quality than animal protein. One industry website advises, "All proteins are not created equal. High-quality animal protein . . . helps fuel a healthy, active lifestyle."[39] Such claims that animal protein is "high quality" and "healthy" are central to the industry's protein dogma, and for that reason, they merit a closer look.

Chapter 6 contains a detailed look at the effects of animal-based foods on our physical health, and the economic consequences that flow from these health issues. But if you want an appetizer, consider the results of a large number of studies on the effect that animal protein has on cancer growth, discussed in the 2004 book *The China Study*. The main finding from these many studies, according to lead author T. Colin Campbell, is that "nutrients from animal-based foods increased tumor development while nutrients from plant-based foods decreased tumor development."[40] This remarkable set of studies, funded by the National Institutes of Health, the American Cancer Society, and other organizations, lasted more than nineteen years and spawned more than one hundred scientific papers published in peer-reviewed journals.

I asked Gregory Miller of the National Dairy Council about Campbell's finding that animal protein, particularly the protein casein in milk, promotes cancer. According to Miller, who has a PhD in nutrition, Campbell's research shows "if you feed [animals] a good healthy diet with a high-quality protein, the cancer thrives, and if you feed them a diet that's not as good, it doesn't thrive. It's about good nutrition." In other words, animal protein promotes cancer because of its high quality, and plant protein *does not* promote cancer because of its poor quality. If this isn't enough to make you curious about the so-called "quality" difference between animal and plant foods, there's even more to it. For additional discussion of this and other topics in meat and nutrition, *see* Appendix A.

The Meaning of *Is*

With an avalanche of emerging research showing that Americans routinely consume unhealthy amounts of meats and other animal foods, animal agribusiness is under pressure. Remember Bill Clinton philosophically—and awkwardly—questioning the meaning of *is* in front of TV cameras? Like Clinton and so many others do when the heat is publicly on, the animal food industry painstakingly contorts its words to sidestep mounting problems.

To be fair, industrial communicators may not intend to mislead or deceive consumers. But frequently their messages are based on data or research that is flawed, out of date, or easily manipulated. Let's reflect on the shaky origins of much of our current thinking on protein consumption. America's obsession with animal protein has its roots in the work of an obscure German scientist who died a century ago. Carl von Voit, a chemistry professor at the University of Munich in the late 1800s, promoted a set of dubious protein recommendations that profoundly influenced the dietary attitudes and habits of Western civilization. Voit based his advice on the amount of protein eaten by men in various professions, such as "well-paid mechanics" and those "at hard work."[41]

But the key flaw in Voit's scholarship was his assumption that because people *did* consume a certain amount of protein per day, they *should* consume that much. As a result, his conclusions were based not on how much protein was healthy, but simply on how much his subjects ate. As a biographer noted, one of Voit's favorite research subjects was his longtime assistant, "a robust man, blessed with a robust appetite, and for this reason, it is said, Voit's well-known dietary standard for a healthy man at moderate work is generously high."[42]

Voit also opposed plant-based diets because he believed protein was most digestible when obtained from animal sources.[43] But in fact, the opposite is often true. For example, according to UN/WHO, the protein in soy, farina, peanut butter, refined wheat, wheat flour, and wheat gluten is more readily digestible than that in meat or fish.[44]

To understand the kind of diet Voit advocated, consider the 150 grams of daily animal protein he recommended for "hard workers." Relying on the USDA's 19th-century estimates of animal foods' protein content, a hard worker might have eaten half a chicken for breakfast, two half-pound steaks for lunch, and a one-pound slab of cod for dinner. The Voit Standard was attacked by contemporary nutritionists who argued that a healthy portion of daily protein for adult males was 40 grams, not 118 or more.[45] Nevertheless, the Voit Standard found wide adoption and acceptance in the United States and other Western countries where, because people already ate a lot of meat, the high standard fit the existing model like a well-tailored suit.

Although the protein levels first urged by Voit have been revised downward over the years, his superlative emphasis on animal protein—and lots of it—continues to serve as a guiding nutritional principle in the United States. Considering that much of American nutritional policy grew from such shaky foundations, is it any wonder today's consumers are confused about what—and how much—we should eat?

Food for Thought

- The animal food industry engages in an aggressive, systematic messaging campaign designed to boost demand for its products. Because the bare facts are usually damaging and likely to hurt sales, the industry often makes its case with ad hominem attacks, sponsored research studies, and false or misleading messages.

- Objective data frequently refutes and exposes this routine industry whitewashing. Case in point: research published by the USDA and the National Academies establishes that Americans eat much more animal foods than we should and that a plant-based diet supplies ample protein.

- The animal food industry often seeks government support to lend credibility to the false or misleading information it disseminates. For example, although the swine flu started in pigs, industry pressure led the USDA to publicly deny this notion. And while the evidence shows that California cows are anything but happy, the California Department of Food and Agriculture seeks to dispel this idea in order to help dairy farmers sell more product.

Sausage Making and Lawmaking: Influence in the Political Process

The idea for this book grew out of an email exchange I had a number of years ago with the dean of a prominent law school. I sent him a link to a short film showing conditions on factory farms. I had just seen the film for the first time myself and was curious what he would think of the images of animal abuse. He wrote back that the treatment of animals in the video was deplorable—and almost certainly illegal. But here's the twist: because he believed it was illegal, he also judged it atypical and therefore, basically irrelevant. As I explored the subject, I found that many share the learned dean's assumption—including most federal lawmakers. My own congressman, Dana Rohrabacher (R-CA), assured me in response to a similar query that "animal cruelty is not only wrong, but against the law."[1]

In civilized society, it is a basic principle that animal cruelty is illegal. Yet as Ronald Reagan cautioned those who would accept too much on faith: "Trust, but verify." As a lawyer, I was interested in legal support for the principle many of us take for granted—that cruelty to farm animals is, in fact, illegal. What I learned in my research surprised me.

In fact, decades of aggressive lobbying by industry groups have yanked the teeth out of dozens of state and federal laws that once protected both consumers and animals. The concerted legislative effort involves much more than just changing anti-cruelty laws. Those who produce and sell animal foods have also been remarkably successful at passing laws that prevent us from investigating, criticizing, or suing them. Industry leaders even go so far as to assert these measures help consumers. In reality, these laws mainly serve the animal food industry and thus, like so much in meatonomics, are simply about the money. This chapter explores how the forces of meatonomics press state and federal lawmakers across the United States to protect industry, and the consequences for American consumers and farm animals.

How to Get to Washington

If Mr. Smith wanted to go to Washington today, he'd need almost $7 million to get elected. Since 1960, the cost to win US office has increased tenfold (in inflation-adjusted dollars).[2] Not surprisingly, as campaign costs have grown, the sums spent on access to those in office have also increased. Organizations increased the annual amounts they spent to lobby Congress between 1983 and 2010 by a factor of thirty-five, and registered lobbyists have multiplied tenfold since 1976.[3] Advocates for the disenfranchised or unprotected, like the poor, minorities, animals, and the environment, lack the budgets to spend in these lavish ways; nearly all the organizations that spend big money on lobbying are businesses or trade groups that represent businesses. Take an imaginary stroll down Washington DC's K Street, where most of these influence machines are located. Nineteen of the top twenty lobbying spenders in 2010 were business interests who spent as much as $157 million per group that year to sway lawmakers.[4] (The one nonbusiness group in the top twenty was the AARP.)

The animal food industry is just one of many special interests to capitalize on this massive change in spending and influence, but it stands out because its efforts have been particularly successful. Most significantly, costly legal restrictions that once applied to its production units—farm animals—have been largely swept aside. Today, at both the state and federal level, legal protection for farm animals is scant, poorly enforced, and largely irrelevant. To put it bluntly, farm animals "have no legal protection at all," according to David Wolfson and Mariann Sullivan, lawyers who have explored the issue in an article titled "Foxes in the Hen House." "As far as the law is concerned," wrote Wolfson and Sullivan, farm animals "simply do not exist."[5]

If this development surprises you as much as it did me, you may wonder, as I did, what it means and how it happened. The answer starts with one of the fundamental principles of lawmaking—the quid pro quo, or this for that, that pervades the American legislative process at both the state and federal levels. While a legislator's express agreement to sell a vote is illegal, it's completely legal to vote for a measure just because a donor likes it. It's a subtle—and for most critics, frustrating—distinction.

Money Talks

Jesse "Big Daddy" Unruh, the late speaker of the California Assembly and a larger-than-life character, was outspoken on the issue of lobbying and influence. "If you can't drink a lobbyist's whiskey, take his money, sleep with his women, and still vote against him in the morning," said Unruh, "you don't belong in politics." (Big Daddy was clearly a man of principle!) But despite such admonitions, and perhaps even acting with the best of intentions, legislators often respond to contributions by doing what the donors want. It's basic human nature—people have a hard time accepting a gift and not reciprocating. In one study of this phenomenon, those who received a can of Coca-Cola as a gift purchased twice as many raffle tickets from the donor as those who got no gift.[6]

Similarly, research regularly finds that donations to lawmakers make it more likely their votes will support the donor's interests.[7] In one example, members of the House of Representatives who received money from the dairy industry were almost twice as likely to vote for dairy price supports as those who received no money. Even more telling, the greater the amount of dairy money a member got, the more likely her support for the legislation.[8] While some might argue that contributors just seek out like-minded legislators, another study of how Congressional lawmakers respond to cash gifts suggests it's really the money that talks. The report found that, in fact, "changes in contribution levels determine changes in . . . voting behavior."[9]

But it's not just a desire to reciprocate that's at play—there's also a basic drive for self-preservation. As we've seen, it costs real money to get elected. Those who violate the principle of quid pro quo are likely to see their donors turn elsewhere, and that could make it hard to stay in office. Marion Nestle, whose book *Food Politics* explores the food industry's influence over lawmakers and regulators, writes of this economic pressure:

> Given the costs of election campaigns, the lack of public funding for them, and the resistance of Congress to reform campaign finance laws, it is no mystery why legislators might not want to make decisions that displease . . . contributors.[10]

From the donors' perspective, the investment bounty from well-placed gifts can be substantial. One study found that in connection with federal subsidy legislation, a $1 industry donation typically yields a $2,000 return in the form of subsidy payments.[11] It's a rate of return so high, most businesspeople would say it's impossible. So is it any wonder the animal food industry spends more than $100 million yearly paying lobbyists and making strategic donations?[12] A look at the industry's dozens of legislative victories in recent decades shows that this spending is yielding more than just robust subsidies. It's also paying back animal food producers in the form of valuable laws that protect and insulate industry players and lower their costs of doing business.

There Oughta Be a Law—Common Law and Animal Protection

Consider the two possible sources of legal protection for farm animals: published cases forming centuries of common law and statutes passed by Congress or state legislatures.[13]

First, a little history primer. Under the common law *free men* had legal protection from physical abuse, but women, slaves, indentured servants, and animals were mere property and could be beaten or abused as necessary to "correct" them. In one example of the common law's treatment of such items of property, a North Carolina court in 1868 declined to punish a man who beat his wife with "a switch about the size of one of his fingers." Applying the so-called Rule of Thumb, the trial judge held that because the switch was smaller than the man's thumb, the beating was lawful. The appellate court declined to apply the Rule of Thumb but upheld the decision on different grounds, refusing to punish the victim's husband "for moderate correction of her, even if there had been no provocation for it."[14] Needless to say, this isn't the kind of ruling that gives judges a good name.

Farm animals have served as monetary currency for millennia, and in some countries still do. In fact, the word *cattle* and the related term *chattel* (personal property) both have their roots in *capitale*, Latin for "wealth." Because the common law saw animals as mere property, it prohibited abuse by third parties that might hurt their economic value but did not punish abuse on the grounds that it was cruel. "The common law recognized no rights in . . . animals," observed a Mississippi court in the 19th century, "and punished no cruelty to them."[15] Accordingly, our inquiry into farm animals' legal protection must turn from law created by judicial interpretation to those statutes passed by legislators.

Humanity Dick—Statutes and Animal Protection

Dick Martin, a colorful lawyer elected in 1776 at age twenty-two to the Irish Parliament and later to the British Parliament, was known as much for his odd antics as for his animal advocacy. He is said to have survived two shipwrecks and over a hundred duels. In one typically dramatic episode, he won a judgment of £10,000 against his wife's lover and then threw the cash from a coach window as he drove through London. If Dos Equis was promoting the Most Interesting Man in the World back then, Martin might have been a candidate.

Martin was also one of the first to respond to the common law's failure to address animal cruelty. He helped found the first Society for Prevention of Cruelty to Animals (now with "Royal" before its name and an annual budget of $200 million). Sometimes called "Hairtrigger Dick" for his dueling activities, and sometimes "Humanity Dick" for his love of animals, he drew inspiration from both pursuits with the remark, "Sir, an ox cannot hold a pistol."

Martin was ridiculed in the press for his animal advocacy and caricatured in drawings that showed him with a donkey's ears. Nevertheless, he authored the most important animal cruelty statute of his time: the Cruel and Improper Treatment of Cattle Act of 1822 (also known as Martin's Act). The act made it a crime to "cruelly beat, abuse or ill-treat" any horse, mule, ox, sheep, or cattle, and it was the first in a wave of anticruelty legislation in Britain and the United States that continued through the 19th century.

In fact, the stated purpose of early anticruelty legislation was to criminalize the abuse of farm animals.[16] Over the course of the 19th century, the scope of legislative anticruelty protection grew wider and expanded beyond farm animals. In 1865, Vermont adopted anticruelty legislation applicable not only to livestock but also to "other animals," and in 1867 New York passed an anticruelty statute that applied to "any living creature."

Legislatively, things started looking good for farm animals after the Civil War. But plot on a graph the level of legislative protection enjoyed by farm animals over the last several millennia, and you'd see a distorted bell curve, sort of like a brontosaurus, very skinny at the left end, peaking in the late 19th century, and declining to roughly zero today. In other words, legislation prohibiting cruelty to farm animals reached a high-water mark in the late 1800s and early 1900s, began declining, and is now almost nonexistent. While legislation safeguarding so many other disenfranchised and exploited groups— children, women, workers, minorities, the disabled, and the elderly— has improved over the last century, how did farm animals come to lose the little legislative protection they once had?

Customary Farming Exemptions

Every US state has a statute prohibiting cruelty to animals. But in response to industry lobbying, most states have adopted an exception to their anticruelty statutes for farm animals. Legal commentators Wolfson and Sullivan, who in 2004 noted the rise of these exceptions, termed them "Customary Farming Exemptions," or CFEs. Connecticut, for example, an early anticruelty pioneer, passed legislation prohibiting cruelty to "any animal" in 1854. But 142 years later, the Nutmeg State exempted all farm animals from the law's protection by adopting a CFE that made "maliciously and intentionally maiming, mutilating, torturing, wounding, or killing an animal" lawful, provided the act is done "while following generally accepted agricultural practices."[17] Three-quarters of US states have some form of CFE, and most of those exemptions were adopted in the past few decades.[18]

Customary farming exemptions essentially remove from lawmakers the authority to decide what constitutes cruelty to farm animals, instead turning that decision over to the farmers themselves. For example, if one farmer decides it would be expedient to chop off all or part of an unanesthetized animal's body part, like an ear, tail, beak, or genitals, and others in the industry follow suit, that procedure becomes customary and protected as a CFE. Thereafter, those who engage in what would otherwise be criminal animal cruelty are exempt from prosecution. As politician and diplomat Andrew Young wryly noted, "Nothing is illegal if one hundred businessmen decide to do it."

A few of the many common farming practices now deemed lawful under most states' anticruelty statutes include:

- Crushing or severing the testicles of unanesthetized animals.

- Slaughtering chickens while they are awake and alert.

- Killing unwanted male chicks or spent laying hens by suffocation, starvation, or disposal in a garbage can or wood chipper.

These practices are not driven by a sadistic urge to be cruel but by a desire to minimize costs. Meat producers prefer to keep their animals alive until they're ready for slaughter, but from a purely economic perspective, it doesn't pay to worry about an animal's pain or suffering in the meantime. It would cost about $0.25 to anesthetize an animal before castrating him.[19] But unless legislatively mandated, the cost of such humane measures to any producer choosing to adopt them would mean lower profits and a loss of competitiveness. As the National Pork Board explained in its 2003 Swine Care Handbook, "Consumers must not expect individual farmers to undertake practices that will make them uncompetitive in the marketplace. Livestock producers will do what is necessary to compete, or else they will not be livestock producers for very long."[20]

A number of countries have outlawed many of these so-called common farming practices. The European Union has banned battery cages, veal crates, gestation crates, and the slaughter of conscious poultry. California has also banned (with effect in 2015) some of these practices, and a few other states have, in part, followed suit. Yet the general rule in the United States is that these practices are not only customary and legal, but also necessary to compete. Because 99 percent of the animals raised for food in this country are grown in factory farms where these practices are commonplace, virtually all of the animal products consumed in the United States come from animals treated in these ways.

By the way, if the way your meat is produced bothers you, you might want to think twice before sharing your opinion too widely (or make sure your opinion is solidly grounded in fact). A growing number of states—thirteen at last count—have laws forbidding "food disparagement." That's right—if you thought libel and slander cases just featured movie stars suing tabloids for gossiping about their love lives, think again. It's unlawful to defame food in a quarter of US states. On her show in 1996, Oprah Winfrey said of the threat of Mad Cow disease: "It has just stopped me cold from eating another burger! I'm stopped!"[21] When Texas cattle ranchers sued her for food disparagement, seeking $12 million in damages, she prevailed only after spending years in litigation and an estimated $1 million in legal fees.

Enforcement Challenges

Even where state anticruelty laws aren't subject to customary farming exemptions, these laws are difficult to enforce for a number of reasons. First, as with most crimes, violators of anticruelty laws can generally be prosecuted only if they act with actual criminal intent to commit a cruel act—simply being neglectful (that is, negligent) is not sufficient.

Second, in most cases, only government prosecutors, not private citizens, can enforce criminal laws. In one notorious case, a San Diego district attorney declined to prosecute chicken farmers who killed thirty thousand "squirming" hens in a wood chipper because they could no longer lay eggs.[22] The DA found no evidence the farmers intended to act cruelly.

Third, violations of anticruelty laws typically occur behind closed doors on private property. Because law enforcement officers are rarely both willing and able to obtain a search warrant (which requires convincing a judge of probable cause that the target has committed a crime), the only way to obtain evidence of wrongdoing is through an undercover investigation. It's hard enough to place an investigator inside a factory farm in a state where it's legal to do so. But a number of states have made it even harder. Under pressure from meat and dairy producers dogged by a constant stream of undercover footage showing inhumane and unhealthy conditions at their facilities, at least seven states have made it illegal to enter an animal facility under false pretenses or to film or take photographs on a factory farm without permission.[23] Since few investigators are eager to spend a year in jail, these so-called ag-gag laws make undercover work on farms in gagged states costly and impractical.

Proponents of ag-gag laws are remarkably candid about the laws' main purpose: to protect industry. "Agriculture is one of our most important industries," says Senator Joe Seng (D-IA), who introduced that state's ag-gag law. "It's sort of a protection mechanism."[24] Protection, that is, from the harmful economic effects that often result when conditions on factory farms are made public. As one rancher wrote, undercover videos "wreak havoc on the agriculture industry, which usually results in litigation, loss of jobs and a direct shot at the markets."[25]

Industrial animal farmers are as tenacious as the proverbial dog with a bone, and they've made ag-gag laws one of their main areas of focus. The lawmaking landscape changed so much while I worked on this book that I rewrote this section four times. The legislatures of Florida and New York rejected ag-gag laws in 2011 and 2012. Iowa's legislature rejected an ag-gag bill in 2011 but passed a slightly different version in 2012. When this book went to press, ag-gag laws were pending in at least six states (Arkansas, California, Indiana, Nebraska, Pennsylvania, and Tennessee) and were expected to be introduced or reintroduced in another six (Minnesota, New Hampshire, New Mexico, North Carolina, Vermont, and Wyoming). The animal food industry's enormous economic and political muscle is all the more impressive when you consider that, according to surveys, laws prohibiting unauthorized filming in factory farms are highly unpopular. In Iowa, for example, just 21 percent of voters approved the ag-gag bill that was defeated there when first introduced.[26]

Wrist Slapping

Another enforcement problem with state cruelty laws lies in the feeble and generally ineffective penalties imposed on violators. Consider the case of Daniel Clark, a pig farmer who, in the winter of 2009, abandoned 832 pigs to die in an unheated barn in Pennsylvania. The dead animals were discovered nine months later in a "mummified" condition.[27] Clark was charged with 832 counts of animal cruelty, but he pled guilty to only ten counts and was fined $2,500. The remaining 822 counts were dismissed, and he received no jail time.

Compare Clark's penalty to that of C. C. Baird, an Arkansas dog broker whose routine abuse of dogs bound for research facilities was documented in the HBO documentary *Dealing Dogs*. Baird, who had 750 dogs in his facility, was charged with laundering money and violating the federal Animal Welfare Act. In addition to being sentenced to six months' home detention and three years' probation, Baird was fined $262,700 and permanently stripped of his dealer's license. He forfeited another $200,000 plus seven hundred acres of land worth an estimated $1.1 million. He also paid $42,000 to animal rescue groups who helped care for and re-home the animals abused in his facility. In all, Baird paid monetary penalties totaling more than $1.6 million.

The difference in punishments accorded to Clark and Baird for their mistreatment of roughly the same number of animals shows the enormous difference in our legal system's view of dogs and pigs. Dogs are companion animals we see every day, whose individual owners advocate for them, and they're protected under a variety of state and federal laws (although, many would argue, still not adequately). Pigs are farm animals we rarely see or think about, whose quality of life runs contrary to the profit motive of the $52 billion pork indus-try. Yet pigs are smarter than dogs and, like dogs, seek mental and physical stimulus and want to engage in various natural behaviors.[28] Is it appropriate that the law allows farmers to abuse pigs with virtual impunity?

Cheeseburger Laws

Hundreds of scientific studies published in respected, peer-reviewed journals link meat and dairy to obesity and disease (*see* chapter 6). Having successfully used comparable studies to extract cash from those who make cigarettes, asbestos, and other harmful products, plaintiffs' lawyers have begun testing the waters to see whether obe-sity studies can be used to impose liability on the food industry. So far the Magic 8-Ball says, "Ask again later." Although a court dismissed the first lawsuit against McDonald's for obesity, what's frivolous to one judge might be meritorious to another—or what lacks merit today could yield a million-dollar judgment in a few years (especially if evidence of industry wrongdoing comes to light). In response to this threat, the National Restaurant Association has led food industry groups in a campaign to beef up their immunity to damages from such lawsuits by urging lawmakers, often successfully, to pass what they call cheeseburger laws.

Before considering these indigestion-causing laws, it's enlighten-ing to look at how such legislation gets started. The American Leg-islative Exchange Council (ALEC) is a conservative forum where corporations and state lawmakers meet and draft corporate-friendly legislation. With the unambiguous motto "Limited Government, Free Markets, Federalism," ALEC is the nation's leading incubator of bills that seek to protect corporations by limiting damages in lawsuits involving issues like product liability, premises liability, and asbestos liability.

Corporations join ALEC for access to state legislators. Legislators join for perks like free trips, dinners, and theater tickets, all paid for by corporate sponsors' generous dues.[29] Lawmakers introduce about a thousand of ALEC's model bills each year in their home states, and about two hundred pass. ALEC flew largely under the radar for decades but was recently in the spotlight when it was revealed that the organization is behind minority-unfriendly voter ID laws that passed in thirty-four states. After a civil rights group disclosed this legislative agenda, a number of ALEC's corporate sponsors, including McDonald's, Wendy's, Coca-Cola, PepsiCo, Yum! Brands, Procter & Gamble, and Intuit, stopped funding the organization.[30]

Cheeseburger laws are typically based on a model ALEC statute titled the Commonsense Consumption Act. Although cheeseburger bills died four times in Congress, twenty-four state legislatures have rallied to the food industry's cries for help.[31] The laws exempt food producers, promoters, and sellers from liability to a plaintiff who claims that long-term consumption of food led to obesity or obesity-related disease.

Cheeseburger laws' proponents say the legislation stands for personal responsibility. "I recognize that obesity is a serious problem in America, but suing the people who produce and sell food is not going to solve this problem," says US Senator Mitch McConnell (R-KY), who sponsored a cheeseburger bill that died in the Senate. "Americans need to take greater care in what—and how much—they eat."[32] But the laws' critics say courts already have a simple mechanism for dealing with frivolous suits—dismissing them—and that exempting industry groups from liability could preclude plaintiffs from recovering for real wrongdoing.[33]

The Green Scare

To recap to this point, animal food producers or restaurateurs have been largely responsible for eliminating farm animals' protection from inhumane treatment in most US states, barring people from suing over obesity and related illnesses in almost half of states, making it illegal to criticize food in a quarter of states, and in one in seven states, preventing undercover investigations into the conditions that might tempt one to engage in such criticism. If any doubt remains as to the clout or determination of those in the industry who regularly convince state legislatures to pass such laws, consider another of the animal food industry's campaigns to stifle criticism: animal enterprise terrorism laws.

Brought to you by the same group responsible for cheeseburger laws, ALEC, a particularly bizarre class of new law has been hitting the statute books in the past decade: so-called ecoterrorism laws. Typically based on a model ALEC statute titled the Animal and Ecological Terrorism Act, these laws provide enhanced penalties for disrupting the operations of an animal enterprise.[34] Criminal conduct that would normally constitute basic trespass, vandalism, or theft is punished more severely when the target is a farm, restaurant, grocery store, or any other animal-related business. Law professor Dara Lovitz writes of this phenomenon:

> The perplexing result is that an arsonist who burned down a building with no political motivation would be tried and sentenced as an arsonist; but if the arsonist burned down the building to protest the abuse of animals therein, s/he could be labeled a terrorist and thrown in prison for exponentially more years.[35]

In a wave dubbed the "Green Scare" because of its similarity to the Communist Red Scare that swept the United States in the 1950s, the federal government and at least thirty-nine US states have adopted ecoterrorism laws in the past decade. Just as the Red Scare was promoted by a handful of fear-mongering individuals, the Green Scare seems to have originated with a handful of special interest groups who profit from animals and rely on their continued ability to do so.[36] With little evidence of actual terrorist activity to point to, these groups nevertheless label animal and environmental activists as "terrorists" and promote what they claim is antiterrorism legislation across the United States to protect their business interests.

According to journalist Will Potter, whose book *Green Is the New Red* explores the Green Scare phenomenon, "'Terrorist' has replaced 'communist' as the most powerful word in our language."[37] *Terrorism* is an emotionally and politically charged term without an accepted definition—in fact, it's officially defined in more than one hundred ways.[38] The terrorist label is often applied by governments seeking to marginalize or delegitimize disfavored groups, leading a *Reuters* editor to instruct writers to avoid using the term: "One man's terrorist is another man's freedom fighter."[39] Under most conventional definitions, terrorism is the use of violence to achieve a political or ideological goal.[40] But this begs the questions: What is violence? Is it violent to burn down an empty building? Is it violent for chanting activists to protest in front of a home? Activists involved in both of these scenarios have been charged with felonies under the federal Animal Enterprise Terrorism Act, although in each case, no one was injured.[41]

When it pushes ecoterrorism laws, the animal food industry routinely cites the need to protect the safety of our nation's food system. The main premise of this argument is that food terrorists might disrupt factory farms and wreak havoc on the nation's food supply. However, the idea that food terrorists seek to hurt the nation's food supply has little empirical support. There have been only two known acts of food terrorism on US soil: the Rajneeshee cult's infecting Oregon salad bars with salmonella in an attempt to influence a local election in 1984, and a disgruntled laboratory worker's infecting food with shigella in 1996. Neither incident resulted in any deaths, and neither made it into the FBI's list of Major Terrorism Cases.[42] Moreover, as animal food production is spread throughout more than 18,800 separate facilities around the country, a terrorist group seeking to disrupt the nation's food supply by targeting individual facilities would have to be remarkably determined and coordinated to have any meaningful effect.

In fact, experience shows that food safety is actually much better served by allowing exactly the kind of third-party monitoring that the food industry seeks to deter. After undercover investigators from the Humane Society of the United States discovered diseased cows being dragged to slaughter in 2007, food safety concerns led to the largest meat recall in US history. When legislators and law enforcement target animal activism, they chill this kind of investigation and discourage the kind of healthy inquiry that benefits consumers. Mike

German, a former FBI agent who has written about his career infiltrating terrorist groups, says about ecoterrorism legislation:

> To create a law that protects one particular industry smacks of undue influence and seems to selectively target individuals with one particular ideology for prosecution. Why does an "animal enterprise" deserve more legal protection than another business? Why protect a butcher but not a baker?[43]

That's easy: bakers don't spend $138 million yearly lobbying lawmakers.[44]

Let's Get Federal

When state laws fail, the federal government often comes to the rescue. In fact, when times call for action to protect the nation's voiceless or disenfranchised, it's usually the federal government, not a state government, that takes action.[45] We look to the federal government in hard cases because it is better suited than states to take difficult action. State lawmakers and judges are often too regionally biased, unprofessional, or uncommitted to take important action. Many states pay lawmakers little, resulting in state legislatures filled with part-timers.[46] Further, federal judges are appointed for life and are thus insulated from popular pressure, while state judges are generally elected to multiyear terms and are sensitive to the need to win reelection. That's why it was federal, not state judges, who were responsible for integrating the South in the 1960s and '70s.

With all these reasons to count on the federal government, you might expect it to be the main bulwark to shield farm animals from cruelty. If so, you'd be disappointed. For starters, the primary federal legislation protecting animals, the Animal Welfare Act, doesn't apply to farm animals—so we can ignore it.

The Twenty-Eight-Hour Law

Two federal laws do purport to require humane treatment of farm animals, although as we'll see, these laws have more symbolic than actual value. To begin with, the Twenty-Eight-Hour Law requires that animals being transported across state lines for more than twenty-eight hours must be provided with food, water, and rest. If this law sounds helpful, consider that in most humans' lifetimes, unless we're stranded on a desert island or trapped in an elevator, we will *never* go without food, water, and rest for twenty-eight hours. The European Union, in fact, imposes transport time limits shorter than ours for all animals, and much shorter for cattle, sheep, goats and very young animals.[47]

Notwithstanding that twenty-eight hours is a long time to go without basic comforts, another limitation of the Twenty-Eight-Hour Law is that it doesn't apply to chickens or turkeys—together accounting for 98 percent of all land animals killed for food in the United States. The law also contains loopholes big enough to drive a cattle truck through. The twenty-eight-hour travel limit applies only to interstate trips, so intrastate travel is exempt—even though animals might spend days traveling eight hundred miles or more in a big state like Texas or California. Additionally, the law can simply be ignored when necessary to overcome "unavoidable causes that could not have been anticipated or avoided when being careful."[48] The twenty-eight-hour limit can also be easily extended to thirty-six hours if the animals' owner makes such a request in writing to the carrier.[49] Further, only those who "knowingly and willfully" violate the law are subject to penalties; those whose "mere negligence" results in death, pain, or suffering to animals don't violate the law.[50] Then there's the maximum penalty for violations: $650. The threat of being forced to pay less than the value of one beef steer is unlikely to influence the behavior of a commercial animal transporter.

It should come as no surprise that a law as toothless as the Twenty-Eight-Hour Law is rarely enforced—after all, what's the point of wasting thousands of dollars of government resources to fine someone $650? In the last fifty years, a grand total of zero federal decisions show a government agency trying to enforce the law. And in 2009, a Freedom of Information Act Request to the USDA and the US Department of Justice revealed that neither organization had investigated or prosecuted anyone under the law in the previous five decades.[51] Given the law's applicability to only 2 percent of farm animals, its massive loopholes, its nominal penalty, and the fact that it is never enforced, we can dismiss the Twenty-Eight-Hour Law as purely symbolic.

The Humane Methods of Slaughter Act

The other federal law that purports to require humane treatment of farm animals is the Humane Methods of Slaughter Act (HMSA). HMSA requires that all livestock be rendered "insensible to pain" before being butchered. However, as with the Twenty-Eight-Hour Law, HMSA's scope and relevance are extremely limited because the law is subject to massive exceptions and inconsistent enforcement. There are no fines or penalties for its violation. Chickens, turkeys, and fish are excluded from HMSA, although these animals constitute 99 percent of the animals slaughtered for food in this country. And HMSA applies only in the final instant of a farm animal's life—the moment of its slaughter. Except for the largely irrelevant Twenty-Eight-Hour Law, neither HMSA nor any other federal law protects farm animals from cruelty during their lives other than at the moment of their death.

As another exception, animals are not protected by HMSA if killed according to a religious method, such as *kashrut,* which produces kosher meat, or a similar method for *halal* meat. These methods require that the slaughtered animal be conscious at death, which calls for a knife cut to the neck, severing the carotid arteries and causing rapid loss of oxygen to the brain. Not surprisingly, this method of slaughter is controversial. Slaughterhouse designer Temple Grandin said of her first visit to a kosher slaughterhouse:

> Each terrified animal was forced with an electric prod to run into a small stall which had a slick floor on a forty-five degree angle. This caused the animal to slip and fall so that workers could attach the chain to its rear leg in order to raise it into the air. As I watched this nightmare, I thought, "This should not be happening in a civilized society." In my diary I wrote, "If hell exists, I am in it."[52]

But by far the biggest problem in enforcing humane slaughter is that industrial killing methods, driven by considerations of profit rather than animal welfare, make it impractical or impossible for most slaughterhouses to comply with HMSA. Workers on fast-moving butchering lines must stun, kill, and dismember three hundred or more animals per hour or face discipline from their employers, and at that speed, it's impossible to stun or kill every animal properly. As a result, animals are routinely scalded, bled, skinned, dismembered, and/or eviscerated while awake and fully conscious. In 2000, workers at a slaughterhouse run by the nation's largest meat processor, IBP, Inc. (now Tyson Foods) provided numerous statements under oath about IBP's practices. One worker said in representative testimony:

> Thirty percent of the cows are not properly knocked [stunned] and get to the first legger [worker who cuts off animals' feet] alive. . . . Cows have gone alive from the knocker to the sticker to the belly ripper (he cuts the hide down the center of the cow's abdomen) to the tail ripper (he opens the [rectum]) to the first legger (he skins a back leg and then cuts off the foot) to the first butter (he skins from the breast to the belly and a little bit on the back) to the worker who cuts off both front feet. . . . I can tell that these cows are alive because they're holding their heads up and a lot of times they make noise.[53]

The USDA is responsible for enforcing HMSA. The National Joint Council of Food Inspection, the union that represents the USDA's 7,500 meat inspectors, has repeatedly voiced its frustration with USDA policies that make it difficult for inspectors to perform their job duties. Meat inspectors want to do their jobs and monitor slaughterhouses' compliance with HMSA, but they often lack the access, or the support from the USDA, to do so. In 2001, the meat inspectors' union joined a number of other groups to present a petition to the USDA, citing systemic, nationwide problems in enforcing HMSA and demanding that the USDA do a better job of enforcing the law.

Specifically, the petition stated that slaughterhouses across the country were failing to stun animals properly and were dragging and beating conscious animals, and that USDA inspectors were largely powerless to take any action. In other words, these inspectors were being sent into battle without weapons or support. Arthur Hughes, speaking for the meat inspectors' union, cited a lack of both training and access to slaughter areas as the chief problems facing union members. "We are the people who are charged by Congress with enforcing the HMSA, but most of our inspectors have little to no access to those areas of the plants where animals are being handled and slaughtered," Hughes said. "The HMSA is not . . . a priority at all. And the USDA doesn't train us—many of the new inspectors don't even know the HMSA exists."[54]

The following year, Congress also expressed its frustration with the USDA's failure to enforce HMSA. Congress stated in the 2002 farm bill: "It is the sense of the Congress that the Secretary of Agriculture should fully enforce [HMSA]," and "it is the policy of the United States that the slaughtering of livestock . . . shall be carried out only by humane methods, as provided in [HMSA]."[55] The only reason for Congress to take this unusual step—restating the provisions of existing law—was to remind the USDA to enforce the law.

Despite petitions, Congress's admonition, and a regular stream of undercover investigations and newspaper articles documenting problems in the nation's slaughterhouses, the USDA's enforcement of HMSA continues to be grossly inadequate. In the fall of 2007, the Humane Society of the United States concluded a six-week undercover investigation at the Hallmark/Westland Meat Packing Company in Chino, California. The investigation found that although five USDA inspectors were regularly stationed at the plant, diseased and infected immobile "downer" cows were routinely slaughtered and illegally sold as meat for human consumption. Because the cows were unable or unwilling to walk to slaughter, various inhumane techniques were employed to move them. These techniques, all of which violated HMSA, included gouging their eyes with a baton, kicking and beating them, and shooting pressurized water into their nostrils and face. This investigation led to the USDA recalling 143 million pounds of tainted beef—the largest beef recall in history. Several employees at the plant were charged with crimes, and the plant ultimately closed down. Where were the inspectors? Why didn't they report this illegal activity?

Stanley Painter, then head of the meat inspectors' union, was summoned to Congress in 2008 to answer these questions. He testified that a number of problems routinely hamper meat inspection efforts, including funding cuts, inspector shortages, increased workloads, and supervisors' practices of intimidating line inspectors and rewriting their reports to exonerate violators.[56] USDA records indicated the department was short eight hundred inspectors, and a watchdog group found that from 1981 to 2007, the number of inspection personnel per pound of meat inspected dropped by 54 percent.[57]

Factory Problems—Rare or Routine?

When undercover footage emerges to show unsafe or inhumane conditions in meat plants, as it often does, the meat industry is quick to respond. After the massive 2008 beef recall, the American Meat Institute issued a statement which read in part:

> The U.S. meat industry has operated under continuous federal inspection since 1906 when the Federal meat inspection system was created. . . . Such intense oversight is unique to our industry. No other industry in agriculture or in other industries, from health care to auto manufacturing, has inspectors on site at all times. . . . Claims that we are not regulated heavily enough or that inspection oversight is lacking are simply outrageous. . . . We will not let a video from what appears to have been a tragic anomaly stand as the poster child for our industry.[58]

It's accurate that the industry is heavily regulated with respect to the safety, packaging, and labeling of food. But in the area of humane treatment of animals, we've seen that the meat industry is essentially unregulated. There's little relevant state law, and the only relevant federal law, HMSA, is poorly enforced, covers only 1 percent of animals killed for meat, and doesn't apply to animals during any part of their lives except the moment of their death. Thus, when meat producers claim "intense oversight" requires them to treat farm animals humanely, we should be as suspicious as when Richard Nixon told the world he had nothing to do with that little break-in at the Watergate Hotel.

What about the assertion that abusive and unsafe conditions in factory farms are merely a "tragic anomaly"? Nathan Runkle was a teenager when he founded Mercy for Animals (MFA), but with twenty undercover factory farm investigations under his belt in a decade, the battle-tested muckraker is now a veteran in his late twenties. I asked Runkle about the industry assertion that abusive conditions are anomalous. "Every investigation we have conducted has unearthed heart-wrenching cruelty to animals," said Runkle, and "every single MFA undercover investigation, without exception, has been conducted at a facility selected completely at random."

Four of MFA's investigations led to civil or criminal proceedings against animal abusers. Further, a number of the investigations uncovered conditions that are unhealthy for people. In an email to me, Runkle wrote:

> While our investigators are inside these factory farms and slaughterhouses, they often uncover practices that raise serious human health and food safety concerns. On factory farms, for example, it's common to find dead hens left to rot and decompose in cages with birds still laying eggs for human consumption. During our investigation at a poultry slaughterhouse, we found that many sick and injured birds—some with cantaloupe-sized tumors—were being snapped onto the slaughterhouse line for human consumption. In New York's largest dairy factory farm, we found that cows with pus-oozing wounds and infections—many also caked with manure—were still being hooked up for milking.[59]

Anomalies? Animal slaughter takes place in secret, carefully guarded, and locked-down facilities where the public is denied access, in many cases under threat of heightened prosecution pursuant to ag-gag or ecoterrorism laws. Even when inspectors are present, their access to the killing floor is often restricted, and it's generally impossible for them to meaningfully observe the slaughter of three hundred or more animals per hour.[60] Further, it's implausible to think that an overwhelmed slaughterhouse worker required to kill thousands of animals daily would concern himself with the concept of welfare while on the killing floor.

In December 2010, the USDA's Food Safety and Inspection Service announced a number of measures designed to improve the enforcement of humane slaughter under HMSA.[61] Animal protection groups lauded the USDA's adoption of these measures, but so far, there is little evidence that the measures are having any effect. Accordingly, until evidence emerges to the contrary, we must continue to view HMSA as we do other federal and state laws that concern farm animals: as a toothless, largely irrelevant measure of limited scope and poor enforcement that provides little protection for the 30 million US farm animals slaughtered each day. Such laws do, however, protect the animal food industry's sales, and that explains why the industry pushes for them.

Rethinking Corporate Benevolence

When animal food producers spend millions to influence lawmakers on laws they claim will help society, like food safety, it's important to read between the lines. In the real world, corporations exist solely for the benefit of their shareholders. Thus, by law, corporations act only to increase their own profitability, *not* to help society. Railroad tycoon William Vanderbilt expressed this principle with candor when he responded to a reporter's question about whether his railroad kept a particular line open to benefit the public. "The public be damned," said Vanderbilt. "I don't take any stock in this silly nonsense about working for anybody's good but our own, because we are not. When we make a move, we do it because it is in our interest to do so, and not because we expect to do somebody else good. . . . Railroads are not run on sentiment, but on business principles and to pay."[62]

Former US Labor Secretary Robert Reich argues in his book *Supercapitalism* that Americans should not be misled by corporate messages implying that private firms act for the public good. Reich warns:

> Beware of any claim by corporate executives that their company is doing something to advance the "public good" or to fulfill the firm's "social responsibility." Companies are not interested in the public good. It is not their responsibility to be good. They may do good things to improve their brand image, so as to increase sales and profits. They may do profitable things that may happen to have socially beneficial side effects. But they will not do good things because they are considered to be good.[63]

So it goes with ag-gag laws, ecoterrorism laws, cheeseburger laws, food disparagement laws, and anticruelty exemptions. They generally run contrary to consumers' interests in food safety, corporate accountability, the free flow of information, and the humane treatment of animals. At bottom, these many lawmaking efforts merely promote what Thomas Jefferson called "the selfish spirit of commerce."

Food for Thought

- Animal food producers have used their massive economic and political clout to pass scores of state and federal laws protecting their business and boosting their profits. Over the past several decades, a spate of legislation has emasculated anticruelty laws and made it harder to investigate, criticize, or sue factory farm operators.

- The animal food industry doesn't pass laws for public welfare, but for its own good. Despite industry protestations about making food safe and protecting it from terrorists, legislation supported by animal food producers routinely hurts consumers and animals.

- As a result of the recent wave of legislation friendly to animal agribusiness, the 10 billion farm animals raised yearly in the United States have virtually no protection from cruelty or neglect. Further, it's impossible in nearly half the country to sue animal food producers for obesity or related diseases, and many states now provide heightened civil or criminal liability for those who criticize animal foods or investigate factory farms.

4

Regulatory Conflict and Consumer Confusion

It's triple the size of a normal salmon but, according to supporters, tastes the same. Dubbed "AquAdvantage Salmon" by its creator and "Frankenfish" or "Salmonster" by critics, the animal in question is a gene-spliced Atlantic salmon whose family tree includes growth DNA from a king (or Chinook) salmon and antifreeze genes from an ocean pout. The resulting invention, which is both patented and trademarked, grows year-round and reaches slaughter size almost twice as fast as normal salmon. If the US Food and Drug Administration (FDA) green-lights this biological contrivance as expected, the fish will become the first genetically engineered (GE) animal food approved for human consumption in the United States.

This controversial, fabricated fish has been under consideration at the FDA for almost twenty years. When the agency said in 2010 that it expected to approve the fish for Americans to eat, public outcry was so fierce—including opposition from thirty-nine US senators and representatives—that officials decided to table the issue while reviewing a new environmental assessment. Yet the FDA's latest assessment (issued in late 2012) reached the same conclusion, and as of this writing, it seems likely the transgenic fish will get the final go-ahead.

The AquAdvantage Salmon story provides an interesting look at how the FDA must serve two demanding masters: industry and the public. In this chapter, I explore the nature of animal food producers' influence on the FDA and the USDA—the two federal agencies that regulate the safety and labeling of animal foods—and what this influence means for consumers. Alarmingly, as we'll see over and over again in meatonomics, even where regulatory functions as critical as these are concerned, heavy industry involvement means that regulation is mainly about the money.

Who's in Charge?

The FDA and the USDA split their food safety and labeling oversight duties in odd and complex ways. For example, the FDA is responsible for seafood, the USDA for meat and poultry. However, the FDA is also responsible for products containing 3 percent or less meat, which means, for example, that the USDA handles sausage while the FDA handles sausage pizza.

Eggs have their own unique rules. The FDA oversees the safety of whole eggs, while the USDA oversees both the safety of processed eggs and the packaging and labeling of whole eggs. This bizarre division of labor can create oversight gaps when, for example, eggs meet packaging and labeling requirements but are contaminated with salmonella. In 2010, this particular set of circumstances led to the recall of 380 million eggs—enough chicken embryos to stretch halfway around the planet. In the finger-pointing that ensued, each agency claimed it had done exactly what its regulatory mandate demanded. All these warped rules are like having umpires at a baseball game who rule on certain players but not others. It can make the game a little hard to follow.

Labeling GE Foods

AquaBounty Technologies insists its product is "biologically and chemically indistinguishable from the Atlantic salmon," but not everyone feels that way.[1] In an online poll by the *Wall Street Journal,* only one-third of respondents said they would knowingly eat GE salmon.[2] While some argue such consumer attitudes are irrational and superstitious, others believe it's right to beware of GE foods. The American Academy of Environmental Medicine, for example, recommends a moratorium on all GE foods, citing studies that show they pose "serious health risks, including infertility, immune dysregulation, accelerated aging . . . and changes in the liver, kidney, spleen and gastrointestinal system."[3]

The biotech food industry says a GE food is no different from, say, an apple perfected over generations and bred to favor traits like sweetness and crispness. Yet there *are* significant differences, both in engineering process and in potential effects on human health, between stitching genes together to create a GE food and developing a food through non-transgenic methods like grafting, pollination, or selective breeding. GE foods can combine the genes of two completely unrelated organisms, like a fish and a vegetable (yielding, for example, frost-proof tomatoes). By comparison, non-transgenic methods like grafting and pollination only work if the two combined organisms share a common evolutionary origin. The result is that while non-transgenic foods are generally thought safe for human consumption, the jury is still out on GE foods. Some clinical studies suggest that GE foods are safe.[4] Others suggest they are not.[5] In his book (recently turned movie) *Genetic Roulette*, Jeffrey Smith explores dozens of human diseases, including childhood food allergies, that may have their roots in eating GE foods.[6]

It may take years of further testing on human guinea pigs—that is, consumers—to learn whether GE foods are innocuous or harmful. In the meantime, if people could just distinguish between natural and lab-built salmon, they could make an informed decision about which one to buy. But here's where the story starts to get, well, fishy. If AquaBounty's GE salmon hits your local market, it probably won't be labeled differently compared to natural salmon. That's because unlike European Union nations and dozens of other countries, the United States does *not* require that GE foods like transgenic salmon bear any special labeling. The last time the issue came up officially, the FDA said it was:

> not aware of any data or other information that would form a
> basis for concluding that the fact that a food or its ingredients
> was produced using bioengineering is a material fact that must
> be disclosed under [legal requirements that food labeling not
> be misleading]. FDA is therefore reaffirming its decision to not
> require special labeling of all bioengineered foods.[7]

By the way, to put this issue into perspective, two-thirds of all food items in American grocery stores contain unlabeled GE ingredients.[8]

In the spring of 2012, a coalition of consumer groups called Just Label It submitted a petition signed by 1.1 million Americans to the FDA calling for the agency to require labeling of all GE foods. "This petition has nothing to do with whether or not genetically modified foods are dangerous," said Andrew Kimbrell, an attorney for the Center for Food Safety, one of the groups leading the campaign. "We don't label dangerous foods, we take them off the shelves. This petition is about the citizens' right to know what they are eating and whether or not these foods represent a novel change."[9] In light of the FDA's express guidance on the issue, it is not surprising that the agency has, as of this writing, thus far declined to act on the petition. In fact, in a show of shaky math skills, the agency outraged many of the petition's supporters and telegraphed its likely position on the issue by counting the petition's 1 million-plus signatures as only a *single* comment—that is, as just one of 394 letters and other documents containing public feedback that it received on the issue.[10]

AquaBounty's stock shot up 600 percent in 2010 when the FDA said it was leaning toward approving the unlabeled GE salmon. For such corporations, preserving the labeling status quo is an important part of the sales equation. That's because given the high level of consumer concern over GE foods shown in polls, such items are likely to sell better when a minimum of attention is drawn to them. And that's why Monsanto, DuPont, PepsiCo, and other corporations that profit from the continued ability to sell unlabeled GE products spent an estimated $45 million in 2012 to defeat a GE-labeling initiative in California.

The FDA's lack of responsiveness on GE labeling results from a phenomenon which the late Nobel Prize–winning economist George Stigler described as "regulatory capture." Stigler said that in order to restrict competition, industry typically seeks and acquires government regulation "designed and operated primarily for its benefit."[11] Once an industry exerts control over a regulatory agency, the agency is "captured." Stigler's theory has been observed repeatedly in US government agencies that are in some measure controlled by the industries they regulate: the Securities and Exchange Commission, the Federal Communications Commission, the Interstate Commerce Commission, the Nuclear Regulatory Commission, and many others—including, not surprisingly, both the FDA and the USDA. As we'll see through a number of examples, these agencies have been captured by the animal food producers and animal drug makers they purport to regulate. And as a result, when the interests of industry collide with those of consumers, regulatory watchdogs have trouble choosing a master.

With regulatory capture common in so many industries, some may wonder what makes the animal food industry worthy of particular attention. The answer lies in the enormous and disproportionate harmful economic impacts that this sector imposes on society. As we'll see in the book's second half, the meat and dairy industries are among the most heavily subsidized sectors in the United States, beating even the well-supported oil industry by a factor of four. The animal food industry also generates by far the highest level of externalized costs of any industry, imposing almost $2 in expenses on American taxpayers and consumers for every $1 of product sold at retail.

Hiding the Growth Hormone

One illustrative and long-running controversy pitting consumers against industry lies in the FDA's policy on the genetically engineered hormone recombinant bovine somatotropin (rBST), known generally as bovine growth hormone. When administered to lactating cows, this hormone pumps up milk production by 10 percent or more. The FDA approved the use of rBST in 1993, finding that dosing a cow with rBST resulted in "no significant difference" in the cow's milk as compared to the milk of an undosed cow.[12] And because of this purported similarity between dosed and undosed milk, the agency declines to require any special labeling on rBST-treated milk.

But aside from the FDA and the Monsanto Company, which makes the rBST product Posilac, few informed observers actually believe that rBST milk is the same as regular milk. In 2010, a US Court of

Appeals disagreed conclusively with the FDA's view on the subject, holding that "a compositional difference *does* exist between milk from untreated cows and . . . milk from cows treated with rBST."[13] Among other differences, the court cited research linking rBST to insulin-like growth factor 1 (IGF-1), a carcinogen, and studies showing higher counts of somatic cells (pus cells which arise from bacterial infection) in rBST-treated milk.[14] In fact, because of the research showing its negative effects on human health, animal health, and milk quality, rBST is banned in Canada, Australia, New Zealand, Japan, and most of Europe. Yet under the FDA's guidance and oversight, in the United States you're likely not only to consume rBST in your milk, cheese, and ice cream, but to do so unknowingly.

Food Safety, the FDA Way

It's not that the FDA doesn't often *want* to do the right thing. It's just that the agency is regularly blocked and bullied by industry forces. Consider the agency's forty-year struggle to take action concerning three antibiotics whose use to make farm animals grow faster is highly controversial—penicillin, chlortetracycline, and oxytetracycline. In the mid-1950s, the FDA approved this holy trinity of drugs for farm animals at subtherapeutic levels—that is, at doses below the level required to treat illness. Such low doses are intended solely for growth promotion. Since the fifties, the drugs have been routinely given to farm animals to promote growth: penicillin for poultry and pigs, and chlortetracycline and oxytetracycline for poultry, pigs, cattle, and sheep. After approving the drugs, however, the agency grew concerned that their long-term use might pose a health risk.[15] In 1972, a task force convened by the FDA found that subtherapeutic application of animal antibiotics promotes antibiotic-resistant bacteria found on meat products and with increasing prevalence, in humans (*see* chapter 6).[16] Federal law requires that if the FDA finds an animal drug is unsafe, the agency must begin proceedings to withdraw approval of the drug.[17] In 1977, following years of study that included significant industry input, the FDA issued notices finding that the subtherapeutic use of penicillin, chlortetracycline, and oxytetracycline in farm animals was "not shown to be safe."[18] The notices purported to start the statutory process to withdraw FDA approval of use of the three drugs in this fashion.

However, that's where the process both started and ended. The FDA never held any required public hearings, and it never took further action to withdraw approval of the drugs. In fact, in December 2011, the agency simply canceled the drug withdrawal processes it had started in 1977, citing its expectation that industry would voluntarily phase out subtherapeutic animal drug use.[19] A few months later, the FDA issued a report containing nonbinding suggestions to industry. The report *recommended*, but did not require, that antibiotics of medical importance to humans be used in farm animals only to treat illness and only under a veterinarian's supervision.[20]

Notwithstanding the FDA's 1977 warnings that subtherapeutic use of penicillin, chlortetracycline, and oxytetracycline in animals is "not shown to be safe," and despite its recent recommendations to cease such practices, the drugs are still approved to promote growth in farm animals and are still being actively used for that purpose. The only material recent development is that in 2012, a coalition of consumer and scientist groups won a lawsuit forcing the FDA to resume the withdrawal process it started thirty-five years earlier. The FDA appealed the decision, and the case is still being litigated as of this writing. In any event, even if the appeals court requires the agency to resume its withdrawal process, the outcome of the process will depend on the agency's ultimate findings as to the drugs' health risks.

What changed after 1977? Why did the FDA abandon the withdrawal process? And why did it take a lawsuit to try to make the agency resume its statutory duty? What happened is that, unable by themselves to convince FDA officials to halt the withdrawal process, industry lobbyists asked congressional lawmakers to lean on the FDA. From 1978 to 1981, in response to pleas from industry, House and Senate committees three times insisted that the FDA pause its withdrawal process for the three drugs.[21] Despite the seven years of study that preceded the FDA's 1977 withdrawal notices, and despite a significant body of published clinical studies, the congressional committees told the FDA to conduct more research—a handy, euphemistic device for delay (and in this case, abandonment).

Theodore Katz, the federal magistrate who in 2012 required the FDA to resume the withdrawal process, noted this congressional noise and blamed it for sidetracking the agency from its duties in the seventies. "Importantly," wrote Katz, "none of these [congressional] recommendations was adopted by the full House or Senate, and none was passed as law."[22] Remarkably, even as the matter winds its way through the courts, and maybe ultimately through the FDA, the US Department of Health and Human Services (HHS, which oversees the FDA) publicly acknowledges "a preponderance of evidence that the use of antimicrobials in food-producing animals has adverse human consequences."[23] Yet farm animals in the United States continue to be routinely dosed with these and other growth-promoting antibiotics.

USDA to Americans: Avoid Unhealthy Food but Please, Buy More of It

Remember Doctor Dolittle? One of the more interesting animals the fictional doctor met in his worldly travels was the pushmi-pullyu (pronounced "push-me pull-you"). With a gazelle's head at one end and a unicorn's head at the other, the animal's every attempt to move results in a struggle but little real progress. The USDA is a bit like this two-headed beast, pushing itself in one direction while pulling in the other. The persistent problem lies in the agency's conflicting goals, which lead to internal clashes and often, confusing advice for consumers.

One of the USDA's chief aims is to help industry sell food to consumers—in its words, "enhancing economic opportunities for agricultural producers" and "getting products from producers to . . . consumers."[24] But other branches of the USDA are charged with *protecting* consumers by giving nutritional advice, keeping food safe and free of contaminants, and ensuring that food advertising and labeling are accurate. When sales promotion and consumer protection go toe-to-toe, as they often do, the results can be truly bizarre.

Every five years, the USDA and HHS partner to release *Dietary Guidelines for Americans,* a set of nutritional recommendations that form the basis for federal dietary policy. Like a modern-day set of Mosaic commandments, these guidelines govern all federal nutritional programs and dietary guidance publications. They serve as the basis for the USDA's food plate—a colorful image intended to recommend food proportions. In short, the guidelines affect—directly or indirectly—practically every American.

The first clue that the nutritional advice in *Dietary Guidelines for Americans* might be based more on marketing than science lies in the membership of its drafting group. Nine of the thirteen panelists on the latest Dietary Guidelines Advisory Committee are linked to industry, having served as advisors or consultants to corporations such as Dannon, McDonald's, Kellogg's, Tropicana, General Mills, and others.[25] Because of industry influence via such committee membership and elsewhere in the process, the Harvard School of Public Health dismisses the *Dietary Guidelines* as the product of "intense lobbying efforts from a variety of food industries" and criticizes its recommendations as the flawed result of the "tense interplay of science and the powerful food industry."[26]

Not surprisingly for the product of a nutritional committee dominated by industry veterans, the *Dietary Guidelines* contain some notably mixed messages. The advisory committee apparently sought to discourage consumption of unhealthy foods without offending the businesses that produce those foods by couching its recommendations in vague, technical terminology that few laypeople understand. Thus, the report routinely resorts to scientific jargon instead of simple terms like *beef* and *pork* when advising against consumption of animal foods. For example, the guidelines advise us to reduce "cholesterol" and "solid fats," but fail to provide examples of foods that contain these substances (translation: meat, fish, eggs, and dairy).[27] The guidelines also urge us, bewilderingly, to reduce "saturated fatty acids by replacing them with monounsaturated and polyunsaturated fatty acids" (translation: eat fewer animal foods; eat more plant foods). The average American has little idea what these terms mean or the kinds of food associated with them. It's like telling a child, "You're prohibited from prevaricating," instead of just saying, "Don't lie." This drafting method is particularly strange in light of the guidelines' use of concrete examples in many areas to tell us the foods we *should* eat, such as "vegetables, beans, and peas."[28]

The confusing language in the *Dietary Guidelines* led the Physicians' Committee for Responsible Medicine (PCRM) to file a lawsuit against USDA and HHS in 2011.[29] The lawsuit charges that the agencies violated federal statutes requiring them to publish "nutritional and dietary information and guidelines for the general public" and to base their dietary guidelines on "the preponderance of the scientific and medical knowledge which is current at the time the report is prepared."[30] At this writing, PCRM's lawsuit was still pending.

Just Say Yes to Cheese

The USDA's messaging on cheese provides another remarkable example of how internal conflict can lead to consumer confusion. The agency's nutritional arm urges us to reduce saturated fat intake by eating less cheese, telling us, for example, to "ask for . . . half the cheese" on pizza.[31] At the same time, the USDA's promotional arm aggressively tries to sell us more cheese. Dairy Management, a checkoff-funded group created by USDA delegates in 1995 to promote dairy sales, has a budget of $136 million and 162 employees. The USDA appoints some of its board members, approves its marketing campaigns and major contracts, and reports to Congress on its work.[32] The USDA also provides funds to the group from time to time, including $5 million in 2009.[33] In 2010, Dairy Management teamed with Domino's Pizza to develop and promote a pizza with 40 percent more cheese—the "Wisconsin"—with six cheeses on top and two more in the crust. Just half of a modest, twelve-inch Wisconsin pizza contains 86 ounces of saturated fat. That's 20 percent more than the USDA's recommended daily saturated fat maximum of 70 ounces—and the pizza is just one meal.

Don't expect the USDA's nutritional team to win this battle anytime soon: with a 2012 budget of only $13 million, the USDA's Center for Nutrition Policy and Promotion is poorly equipped to compete in the battle for consumers' hearts and minds—and arteries. As we've seen above, the National Dairy Promotion and Research Board, marketing mouthpiece of the USDA and parent of Dairy Management, boasts a financial arsenal of $389 million in annual checkoff funding.[34]

"Don't tell me where your priorities are," said educator James W. Frick. "Show me where you spend your money, and I'll tell you what they are." The USDA's priorities are complicated, it seems. The agency's inherent conflict of interest in nutritional matters led former US Senator Peter Fitzgerald (R-IL) to propose in 2003 that responsibility for nutritional advice should reside solely with HHS. "The USDA food pyramid [precursor of the food plate] probably has more to do with diabetes and obesity than Krispy Kremes," Fitzgerald said. He unsuccessfully urged his colleagues to strip the USDA of responsibility for dietary advice and leave it to pursue its "main mission . . . to promote the sale of agricultural products."[35]

Arsenic in the Meat

As another symptom of industry influence, the USDA is sometimes as hard-pressed to protect food from contaminants as it is to promote healthy eating practices. In response to a 1993 *E. coli* outbreak at Jack in the Box restaurants, the agency implemented a new meat inspection method known as Hazard Analysis Critical Control Points (HACCP). HACCP changed the focus of inspection to critical control points in meat plants, but in response to heavy industry pressure, it also transferred many inspection duties to the meat industry. HACCP also allowed plants to build walls blocking USDA inspectors from seeing the slaughter area.

The USDA's watchdog arm, the Office of Inspector General (OIG), has noted numerous problems with HACCP. For example, in one report, the OIG found that large meat producers, who have delegated to themselves the authority to determine whether the presence of *E. coli* in their meat is a hazard "reasonably likely to occur," are routinely unsuccessful at making this determination.[36] The OIG found that *E. coli* is typically found at meat plants only by "happenstance," and that meat producers are largely ineffective at self-testing for its presence.[37] Another OIG report noted that inspectors receive inadequate training and oversight and are too few in number to conduct inspections properly. (There are 7,500 USDA food safety inspectors, but with 30 million land animals killed in the United States each day, they're spread a little thin.) This report also found "inherent vulnerability to humane handling violations" because of the lack of continuous inspector surveillance.[38]

Another result of industry's influence on the USDA is in the department's lack of authority to issue a mandatory recall of meat—even when it's contaminated with *E. coli*, pesticides, heavy metals, or veterinary drugs. All the USDA can do is *recommend* that meat producers voluntarily initiate a recall on their own. That's right—the main US regulatory agency for animal food safety lacks the ability to recall tainted food and must rely on companies who sell dangerous food to issue recalls at their sole discretion. It's as if the FAA lacked real authority to direct air traffic and could only politely suggest that pilots stay out of one another's way.

The threat of a lawsuit may encourage some producers to voluntarily recall contaminated meat. But this threat lacks the force and urgency of a mandatory recall, and in any event, lawsuits take years and require plaintiffs to prove both causality and damages—something not everyone has the time or money to do. A 2010 OIG report found that the USDA's inability to recall meat, coupled with a lack of processes to detect and determine safe levels of chemical residues in meat, led to quantities of meat entering the food supply containing pesticides, drugs, and heavy metals like arsenic.[39] These substances can cause a variety of human illnesses, including stomach, nerve, and skin diseases.

What's in a Name?

As we saw with the FDA, food labeling is an area where agency policy can have an enormous effect on consumers' ability to make informed choices. At the USDA, industry influence has led to watered-down labeling standards, and as a result, food labels are often less than forthcoming. For example, the USDA is charged with enforcing correct use of the term *organic,* but lobbyists have succeeded in convincing both

Congress and the USDA to adopt numerous loopholes in the term's legal definition. Foods can now contain any of nearly three hundred nonorganic substances and still be labeled "USDA Organic." These include a variety of synthetic substances, such as artificial colorings, flavorings, starches, casings, gelatin, and hops.

Industry players sometimes use their influence to bypass inconvenient labeling restrictions altogether. Consider an example from the organic baby food market. USDA staff members determined that baby formula containing synthetic fatty acids could not use the term *organic.* Among other concerns, the synthetic fats are often produced using hexane, a neurotoxin. However, after industry lobbyist William Friedman contacted USDA Deputy Administrator Barbara Robinson to argue his clients' position, Robinson overruled her staff and allowed the baby food with the synthetic ingredients to receive the coveted organic designation.[40] In an interview with reporters, Robinson defended her actions by explaining the USDA's main purpose is to "grow the industry."[41] At the time of this writing, Robinson held a management position in the USDA's National Organic Program.[42]

Another compromised component of the organic program is the access-to-pasture requirement. To qualify for the organic label, products of ruminants (i.e., cud chewers) like cattle, sheep, and goats must come from animals who have "access to pasture" during their lives. This phrase remained undefined for nearly a decade, leading to widely disparate interpretations and little oversight. For meat and dairy producers, providing access to pasture is more costly than simply sticking animals in the small, muddy feedlot that is emblematic of factory farming. Animal agribusiness naturally sought to minimize the length of time that ruminants must spend in pasture in order to be deemed organic, and they succeeded.

In 2010, the USDA issued regulations defining the access-to-pasture requirement. In order to qualify as organic, ruminants must now spend only one-third of each year in pasture, and only one-third of their food must be pasture based (i.e., grasses). Nevertheless, these animals may still be fed significant quantities of foods they don't naturally eat, and they may be denied access to pasture for two-thirds of their lives.

Behind the Revolving Door

In the animal food industry's toolbox of regulatory influence, one handy device is the ability to pick up the phone and call former industry insiders in government jobs. The Monsanto Company patents and sells 90 percent of its GE crop seeds in the United States, and the crops from these seeds yield most of the animal feed supplied to farm animals in the United States.[43] Before Monsanto got into GE seeds, its other products included Agent Orange, polychlorinated biphenyls (PCBs), dichlorodiphenyltrichloroethane (DDT), and, as we've seen, the bovine growth hormone rBST. A number of Monsanto's thirty thousand employees have strutted right through the revolving door from business into federal government, snagging influential posts. Monsanto expatriates include a US Supreme Court justice, a US trade representative, US secretaries of commerce and defense, an EPA deputy administrator, and FDA leaders in the posts of branch chief, deputy commissioner for foods, and deputy commissioner for policy.[44]

The USDA's playbill also includes a star-studded cast of industry veterans in high places. According to a 2004 report, a dozen of the USDA's top-ranking officials during the George W. Bush administration previously worked in the agricultural industry.[45] These included the agency's secretary, her chief of staff and deputy chief of staff, a deputy secretary, three undersecretaries, and three deputy undersecretaries. The officials who left the Heartland for stints in the Beltway came from groups like the National Cattlemen's Beef Association, the International Dairy Foods Association, the National Cheese Institute, the American Butter Institute, and a pork plant.[46] Eric Schlosser, author of *Fast Food Nation,* wrote of the USDA in a 2004 *New York Times* article, "You'd have a hard time finding a federal agency more completely dominated by the industry it was created to regulate."[47]

Of course, some agency personnel change when presidents change, but the basic principle of industry control doesn't change much. Obama's appointee to head the USDA, Tom Vilsack, was said by the

Organic Consumers Association to be a "shill for agribusiness biotech giants like Monsanto."[48] While governor of Iowa, Vilsack was regularly invited to fly in Monsanto's corporate jet and was named Governor of the Year by the Monsanto-dominated Biotechnology Industry Organization.[49] As secretary of the USDA, Vilsack returned the favor by approving the use of Monsanto's genetically engineered alfalfa as feed for cattle. In a controversial move that one commentator called a "stunning reversal," Vilsack set aside a carefully worked compromise applauded by food industry critics like Marion Nestle (author of the book *Food Politics*) but opposed by Monsanto.[50]

If you haven't noticed it, there's a pattern emerging. Guidance and protection from the FDA and the USDA on animal food matters are often concerned more with helping industry than with guarding Americans' health and safety. Or perhaps to put it more bluntly, as Rory Freedman and Kim Barnouin do in their best-selling volume *Skinny Bitch*, "Governmental agencies don't give a shit about your health."[51] Because agency guidance often reflects little more than industry's desire for booming sales, our choices are frequently—and unwittingly—not in our own best interest. We might choose not to buy rBST-treated milk, if only we could identify it. We might choose not to consume more than our recommended daily maximum of saturated fat in one sitting with a pizza, if we were warned instead of encouraged. Sometimes the complex division of duties between the agencies results in confusion and finger-pointing. Do we deserve better protection from our watchdogs? The short answer is yes, but more on that later.

Food for Thought

- In a classic fox-in-the-henhouse story that might be funny if it weren't so true, the animal food industry has captured the very agencies meant to regulate it. As a result, policy-making at the FDA and the USDA is heavily influenced by producers of meat, fish, eggs, and dairy, and by those who keep animal agribusiness stocked with supplies like drugs, feed crops, and crop seeds.

- Because corporations seek regulation in order to enhance their profitability rather than to benefit society, industry-dominated policy making usually hurts consumers. Among other things, government complicity in misleading or confusing practices diminishes consumers' ability to make informed decisions.

- At the FDA, industry control has resulted in such harmful results as the agency's inability to withdraw dangerous animal drugs and its refusal to require labeling of genetically engineered foods. At the USDA, regulatory capture leads to inconsistent and misleading nutrition recommendations, as when one branch of the agency tells us to eat less cheese while another tells us to eat more. Industry influence also yields watered-down labeling standards and weak enforcement in the use of regulated terms like *organic*.

II

THE HIDDEN COSTS OF MEATONOMICS

5

Feeding at the Subsidy Trough

In 1991, you could walk into any McDonald's in the United States, hand the teenager behind the counter a single dollar bill, and take home a double cheeseburger or a chicken sandwich. Back then, a gallon of gas cost an average of $1.14 in the United States (today it's $3.83), and the federal hourly minimum wage was $4.25 (today it's $7.25). In fact, the purchasing power of a dollar was 70 percent higher two decades ago. In the words of the immortal Yankee Yogi Berra, "A nickel ain't worth a dime anymore."

But here's a surprise. Walk into a McDonald's today and hand one thin dollar to the teenaged descendant of your former cashier. Unless you no longer eat such things, you can still leave with a double cheeseburger or a chicken sandwich. In fact, animal food prices in the United States have been remarkably resistant to the forces of inflation over the past century, falling across the board (in inflation-adjusted terms) while most other consumer goods continue to rise in price.

In the book's second half, I explore why the retail prices of meat and dairy are so low and what these low prices really mean for consumers. The common, knee-jerk reaction to lower prices is "Great! I can buy more!" But in this case, low prices are both cause and effect of a microeconomic system out of whack. In fact, by keeping the system in a perpetual state of disequilibrium, or market failure, the forces of meatonomics create problems that affect almost everyone. For the skeptical, there is a mounting pile of evidence that artificially low prices actually hurt, rather than help, consumers. This chapter introduces some of the hidden costs of meat and dairy production and explores one particularly controversial item: subsidies.

Cheap Meat

Meat and dairy *are* cheap. From 1980 to 2008, the inflation-adjusted prices of ground beef and cheddar cheese fell by 53 and 27 percent, respectively.[1] During roughly the same period, the inflation-adjusted prices of fruits and vegetables rose by 46 and 41 percent, respectively.[2] The result is that in contrast to their relative prices three decades ago, today a dollar buys three times the ground beef compared to vegetables that it once did.

The main reason for the steep drop in meat and dairy prices is that producers have made animal agriculture practices vastly more efficient. Back in the day, animal farming was heavily land intensive. But the modern shift to high-density, hyper-confinement methods has greatly reduced the need for wide open spaces. Automated processes have reduced labor costs. Consolidation and increased output volumes allow producers to enjoy economies of scale. And animals are now bred to grow larger and reach slaughter weight sooner.

The efficiency gains are noteworthy. Per-hen egg production has doubled in the last century.[3] Per-cow dairy production has tripled.[4] And the average weight of broiler chickens has almost tripled, while the birds' growth rate has more than doubled.[5] On the surface, it looks like innovation is doing what it should, which is to reduce prices. But a closer look at the economic costs of animal foods suggests something more complicated is happening.

Externalities: Looking Outside the Box

If I pay a garbage service to collect my trash, my disposal costs are internalized. Appropriately, because I generated the trash, I pay the collection costs. But if I drive over to the local park at midnight and dump my trash there, I've imposed my disposal costs on others. Such "externalized costs" are those expenses related to producing or consuming a good that are not reflected in the good's price and are instead passed on to third parties. This concept is critical to understanding the economics of animal food production.

As taxpayers, we routinely pick up the check for externalities that everyday transactions generate. Take cigarettes. Because of the costly health problems associated with smoking and exposure to second-hand smoke, the US Centers for Disease Control and Prevention estimate that each pack of cigarettes sold imposes externalized health care costs of $10.47 on Americans.[6] But even with cigarette taxes as high as $5.85 per pack in some areas, governments are far from recovering all of the externalized costs. Cigarette manufacturers, of course, pay almost none of these costs. The result is that the additional price of smoking is borne by many who don't smoke, including taxpayers and those who pay health insurance premiums. In other words, even if you don't smoke, you're writing checks to cover the doctor bills of those who do.

Calculating external costs can be controversial. To summon a well-worn cliché, the devil is in the details. Free market advocates downplay the extent and value of externalities, while those who favor market regulation find costly externalities wherever they turn. For example, imagine that the construction of an oil pipeline harms caribou herds in remote areas. Some might argue that as a non-market animal—that is, a species we don't normally use for food, clothing, or entertainment—these remote herds have no economic value and their decline generates no external costs. Others would say that the amount that animal-friendly humans are willing to pay to protect the caribou is an external cost of building the pipeline.

In the case of animal foods, their low retail prices obscure a significant, measurable set of external costs that make the real costs to society much higher. Many of these expenses, such as taxpayer subsidies, health care costs, and environmental costs, have been extensively researched and documented. When these documented tallies are considered, the true price of a Dollar Menu double cheeseburger turns out to be a lot higher than a buck. Moreover, there is little comfort in the superficially pleasing idea that economic conditions keep prices low for consumers. Anything that *seems* too good to be true probably *is* too good to be true. In fact, low animal food prices are largely illusory because these goods' true costs are shifted to consumers in roundabout ways. Ultimately, the numbers show that the big winners from these heavy external costs are a handful of animal food industry fat cats and the few companies who provide them with supplies like feed, drugs, and equipment.

An Upside-Down Industry

It's common—and usually perfectly legal—for corporations to externalize as much of their costs as possible. And of course, animal agribusiness isn't the only American industry to impose billions in hidden costs on consumers and taxpayers. But this sector is far and away leading the pack in the category, offloading more costs than any other.

Consider electricity generation. That industry is known to impose tremendous external costs on society, mainly in the form of health problems and ecological damage resulting from burning coal and oil. A 2009 study by the National Research Council summed up the quantifiable, externalized costs of US electricity generation and found they total $63 billion yearly (in 2005 dollars).[7] That's a sizable figure, but even adjusting it for inflation ($75 billion), it's less than *one-fifth* of animal foods' measurable external costs.

One way to estimate the total cost to consumers of animal foods is to add external costs to retail prices, since generally, Americans who consume animal foods will incur both the retail prices and the externalized costs.[8] The industry's total annual retail sales are about $251 billion.[9] But that's chump change compared to the total external costs of animal food production, which are shown below to be $414 billion.[10] As Chart 5.1 shows, adding external costs to retail sales yields total consumer costs of about $665 billion.[11] From this perspective, each $1 in retail sales of animal foods generates about $1.70 in external costs. The true cost of a $5 carton of organic eggs is roughly $13. A $10 steak actually costs about $27.

Here's another way to think about the ratio of *prices* (what we pay at the cash register) to *costs* (all relevant expenses—whether paid or not). Over the past century, the large gains in production efficiency that accompanied the rise of industrial agriculture helped drive the retail prices of animal foods lower. The growth rate of chickens doubled while the price per pound fell. However, as Milton Friedman famously observed, "There's no such thing as a free lunch." In fact, the same gains in efficiency that reduce prices also increase externalized costs. Chickens develop faster in part because they're fed growth-promoting antibiotics, but those drugs cause costly antibiotic resistance when they end up in our food and our waterways, making it that much harder for us to fight off a slew of sicknesses—and leading to more time in the doctor's office. Animals packed in factories can be raised more cheaply than those raised on pasture, but separating animals from land means their waste must be collected and stored. Some of this waste ends up in our water supply and generates serious clean-up costs. Like a new car promotion that offers twelve months without payments, the price of factory farming has been deferred or ignored—but by no means eliminated.

CHART 5.1 Total Costs to Consumers of Animal Foods (in billions of dollars)

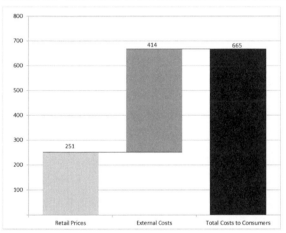

If your diet tends toward the omnivorous, you might feel a sense of satisfaction that the prices you pay for meat and dairy are lower than they would be if external costs were imposed at the cash register. On the other hand, if you don't eat animal foods, you might feel relief that at least you're not participating in the system. But in either case, your wallet or pocketbook is going to feel the hit. That's because, like it or not, even though we don't pay them at the cash register, we all *do* incur the costs of animal food production in one way or another. For example, even if you're lucky enough never to develop cancer, diabetes, or heart disease, you'll still help finance the treatment of those who do (unfortunately, many cases of these three diseases are attributable to consumption of meat, fish, eggs, and dairy).

Regardless of your eating habits, there's no way to avoid this steep burden. Don't eat fish? It doesn't matter—you'll still sustain your share of the expenses of overfishing, algal blooms, and other problems associated with fishing and fish farming. Don't eat meat or dairy? Vegetarians and vegans experience the externalized costs of animal foods at the same rate as the rest of society. Whenever someone buys a Big Mac, herbivores and omnivores alike pick up the tab for the burger's additional, externalized costs. This shifting of costs means each of us—rich or poor, sick or healthy, omnivorous or vegetarian—pays the true costs of these goods, not their producers. Further, contrary to the oft-repeated claim that producers pass low prices on to consumers, the externalization of costs does *not* really save us money because we simply pay the costs in other ways.

Of course, mass-producing just about any food generates external costs, and fruits and vegetables are no exception. Thus, growing crops for people imposes some of the same external costs on the environment that growing feed crops does, such as those arising from the use of pesticides and fertilizers. However, the external costs of growing fruits and vegetables are minuscule compared to those of producing animal foods. Plant-based foods, for example, generate virtually none of the health care costs and far less of the environmental costs that animal foods do.[12] Moreover, government subsidies are heavily skewed toward animal food producers, who receive more than thirty times the financial aid that fruit and vegetable growers do.[13] And like most features of meatonomics, these subsidies work in strange ways and have a number of unexpected consequences.[§]

---

§ Although subsidies are typically not considered an externalized cost, they are included in this category for calculation purposes because, just like external costs, subsidies artificially depress prices and increase consumption, and they impose a large social burden without providing any real benefit to most consumers or taxpayers.

Feeding at the Trough

US agriculture is propped up by an intricate scaffold of government subsidies that would make Rube Goldberg proud. These programs funnel cash and benefits to big farmers in a variety of ways that most nonfarmers have never heard of, including crop insurance, disaster payments, and counter-cyclical payments, to name just a few. Few laypeople understand how these complex programs work, how pervasive they are, or the unexpected places they crop up. I was surprised to learn, for example, of a massive water subsidy program for farmers a few hours from where I live in Southern California. In California's Central Valley, irrigation subsidies let farmers use roughly one-fifth of the state's water and pay only a small fraction of its value—about 2 percent of what Los Angeles residents pay for water.[14]

The farm subsidy system is filled with off-the-wall incongruities, the oddest of which lie in the federal government's conflicted policy agenda. On one hand, as we've seen, the USDA recommends in its *Dietary Guidelines for Americans* that we consume less cholesterol and saturated fat. On the other hand, it heavily subsidizes the foods that are the main sources of these substances—meat, fish, eggs, and dairy. On one hand, the *Dietary Guidelines* recommend that we eat more fruits and vegetables. On the other hand, the government designates these foods as specialty crops that are largely ineligible for subsidies. The net result: nearly two-thirds of government farming support goes to the animal foods that the government suggests we limit, while less than 2 percent goes to the fruits and vegetables it recommends we eat more of.[15] (The USDA, by the way, officially declined to comment on this or any other issues in this book.)

Perhaps even more surprising than our government's muddled messaging on agricultural subsidies is the total dollar value of these subsidies. The USDA will spend $30.8 billion in 2013 supporting US farmers with loans, insurance, research, marketing assistance, cheap water, and other help.[16] State and local governments will contribute another $26.5 billion, mainly in the form of irrigation subsidies.[17] That's a total of about $57.3 billion in annual government subsidies to all segments of US agriculture—more than the entire annual government budget of New Zealand and twice that of the Philippines.

The lion's share of this subsidy largesse supports the animal food industry. Because most US corn and soybean crop is used for livestock feed, and most subsidies help farmers who grow these crops, most of the subsidies to US agriculture ultimately benefit producers of meat, eggs, and dairy. What kind of numbers are we talking about? One study estimates that 63 percent of US subsidies benefit animal food producers.[18] Applying this percentage to the $57.3 billion farm subsidy total, and adding $2.3 billion for fish subsidies (*see* chapter 9), the total of annual subsidies to US producers of animal foods is an estimated $38.4 billion.[19]

Who Gets the Handouts?

These subsidies work in surprising ways, and it's enlightening to consider whom they help and hurt. Supporters of farm subsidies argue that these programs provide a number of benefits, including assisting small farmers, boosting rural development, stabilizing commodity markets, and promoting national food security. However, subsidies typically work in ways far different from those intended.

Let's start with some historical perspective. When the Depression hit in 1929, prices of farm goods fell like corn stalks in a gale. President Franklin Roosevelt responded with New Deal programs, including subsidies, designed to stabilize prices and boost small farmers' incomes. The system worked at first, but as large farms came to dominate the agricultural landscape, less and less of the subsidy funds wound up in the little guys' pockets.

Today, the handouts are aimed at big farming concerns. Taxpayers paid out $161 billion in direct payment farm subsidies between 1995 and 2009, but two-thirds of US farmers didn't receive a cent. The funds mostly went to big corporate players, with one-fifth of recipients grabbing nine-tenths of the cash.[20] If the federal government sent invites to this party, those for small farmers were clearly lost in the mail. The 2013 farm bill (not yet passed as of this writing) seeks to end direct payments, the most controversial subsidy type, shifting the funds to crop insurance instead. But the cash and benefits will still favor the big guys over the little family operations. Moreover, not only do most of the subsidy benefits go to large corporations, but most of the beneficiaries are in the animal food industry.

It may come as little surprise, but the handful of farmers who consistently harvest the most greenbacks from crop subsidies, research shows, are livestock producers. The reason: corn and soybeans are the main items on the menus for livestock, accounting for the majority of feed ingredients in factory farms (where virtually all US farm animals are raised).[21] This makes factory farms the biggest consumers of these subsidized commodities, and they buy most of the corn and soybeans grown in the United States.[22] Fish farms, by the way, also rely heavily on corn and soy for feed.

The expense of feeding hungry animals is a huge part of the cost of doing business in animal agriculture. This food accounts for almost half the costs of raising hogs and nearly two-thirds the costs of producing poultry and eggs.[23] But crop subsidies come to the rescue by helping producers keep their feed costs low. One study found that crop subsidies and the resulting lower feed prices allowed factory farms to decrease their operating costs by as much as 15 percent from 1997 to 2005, saving nearly $35 billion over the period.[24] As these programs favor big operators over small, they not only fail their original purpose of helping small farmers, but they also go a step farther and actually help drive small farmers from the rural landscape.

Buying the Farm in Rural America

The problem isn't just that small farmers don't get their fair share of the subsidy pool. It's also that in light of the unintended effects these handouts have on the market, small farmers are among the biggest losers under subsidy programs. Subsidies distort the normal market forces of supply and demand, making small farmers overproduce even when inefficient to do so. For an example of math that doesn't really add up, consider: when subsidies reduce the market price of corn to levels *below* what it costs to produce, US farmers nevertheless continue to grow it in record amounts.

From 1986 to 2005, the cost to grow corn was higher than its market price in every year but one.[25] Yet because of market incentives to increase production levels, even as subsidy payments tripled during the second half of this period, farmers' net income fell by 15 percent.[26] It's not hard to see why. Say the cost to produce a bushel of corn is $6, but the bushel can only be sold for $5.90. No one's going to grow corn at these numbers. But throw in a $1 per bushel subsidy, and now there's profit of $0.90 per bushel. But as production increases, and the excess supply causes prices fall to $5.50 per bushel, farmers grow more corn to make up the income shortfall. This in turn drives prices even lower. It's a vicious cycle, a perpetual pattern of false cues and financial suffering for many of its participants. As physician Martin Fischer said, "Morphine and state relief are the same. You go dopey, feel better and are worse off."

Supply management—holding goods in reserve—once kept these harmful forces in check. Remember government cheese? In the 1980s, the Reagan administration famously tapped tons of dairy reserves to give processed cheese to the poor. Criticized as the patronizing and unhealthy product of unrealistic and out-of-touch leadership, the orange, five-pound blocks became synonymous with Reaganomics and government double-speak. Still, despite evidence that supply management works, at least from a market perspective, policy makers abandoned that system in the 1990s for a market-oriented approach that deregulates supply. The theory was that with less regulation, farmers would make better choices—like producing less when prices dropped. Unfortunately, it turns out the theory was wrong.

In the world of small farms, subsidies are a poisoned chalice. Depressed commodity prices and smaller margins hurt small farmers' incomes. Not surprisingly, USDA research indicates that even with subsidies, most farm families don't earn enough farm income to support themselves.[27] Most of these families moonlight in off-farm activities, getting second (or third) jobs or engaging in other business activities to make ends meet.[28] This decline in income and lifestyle leads most small farmers and rural Americans to oppose subsidies; nearly two-thirds of Iowans, for example, favor eliminating subsidies.[29]

The economic damage wrought by farm subsidies plays out regularly in rural communities across the country. In a 2007 article titled "Why Our Farm Policy Is Failing," *Time Magazine* described the effect of subsidies on the small town of Randolph, Nebraska:

> Randolph's school district has dwindled from nearly 1,000 students to fewer than 400. It's adopted a four-day week to save money and might switch to eight-man football. The town has lost its Ford, Chevy and Chrysler lots, all its implement dealers and lumber yards, its creamery, jewelry store and movie theater. "The big farmers took over, and it's killed small business," says Paul Loberg, who runs a welding shop off Main Street. "All they need downtown is coffee and beer. They can't buy that by the truckload yet."[30]

Subsidies and the Rise of Factory Farms

Subsidies also cripple small farming towns by promoting the growth of factory farms, or concentrated animal feeding operations (CAFOs). CAFOs are usually bad news for those who live nearby, saddling local regions with pollution-related costs and other burdens. Some locals might accept these consequences if CAFOs gave back to the community by providing jobs or buying goods. But CAFOs hire fewer workers per animal than smaller farms do, and they purchase most supplies from out of the area. Thus, as CAFOs enter a region, they cause declines in local business purchases, infrastructure, and population, turning once-vibrant communities like Randolph into ghost towns.[31] Mike Korth, a farmer in Randolph, received $500,000 in subsidies over ten years but still publicly opposes farm welfare because of the damage it causes by spawning factory farms.[32]

Artificially low feed prices also give CAFOs a huge advantage over the dwindling number of smaller, traditional farms that still grow their own feed. As a result, over the last several decades, CAFOs have consolidated and gotten bigger. At the same time, traditional farms have gone out of business or sold off to their larger competitors, disappearing from the landscape by the millions. The number of US hog farms, for example, shrank by 76 percent from 1982 to 2002 although pork production increased by 10 percent during the period.[33] As this trend continues, fewer small farms survive, and market power concentrates in the hands of fewer and fewer industrial producers. One analysis bolstered this point, finding that four beef packing firms control 84 percent of their market, and four pork packing firms control 64 percent of theirs.[34] Animals, it seems, are not the only ones who are highly concentrated in factory farming.

Worse yet, the larger revenues generated by CAFOs routinely *don't* generate higher tax revenues for the states or localities where they do business. One study looking at state and local tax revenue from swine farms found that among all farm types, CAFOs generate the least tax revenue per pig.[35] Local tax revenues also suffer because lower purchases from local businesses mean decreased sales tax receipts, lower property values in the vicinity of CAFOs cause lower property tax receipts, and property tax exemptions for CAFOs lead to lower tax revenue.[36] The one-two punch of subsidies and CAFOs, it seems, has rural America on the ropes.

Beggaring the Neighbors

In addition to plowing over our own small farmers, US farm subsidies routinely hurt the little guys throughout the rest of the world. Because meatonomics heavily subsidizes feed crops like corn and soybeans, for decades US farmers have sold these crops on world markets at prices well below their cost of production. This is classic dumping—the selling of goods in a foreign market at less than fair value. In the British card game Beggar My Neighbor, one of the goals is to make neighboring players pay stiff penalties. In the real world, when nations dump, they burden their neighbors with genuine penalties in the form of severe economic and social disruption.

In developing countries, where 60 percent of the people still earn their living in farming and sometimes get by on income of $1 per day, the results of dumping can be catastrophic. Small farmers would ordinarily be challenged to compete against a large agribusiness merely because of the economies of scale enjoyed by the larger competitor.

But these small farmers have an even harder time competing when subsidies permit the large business to consistently sell at prices *below its own cost of production.* This slanted playing field drives small farmers out of business and off their land, and it contributes to widespread socioeconomic problems in developing countries.

Dumping by US corporations has had devastating effects on a number of developing nations, including, by way of example, our neighbors to the south. The influx of cheap US corn flattened Mexican corn prices by 70 percent from 1994 to 2003, contributing to a 20 percent drop in Mexico's minimum wage and helping boost the nation's unemployment rate to 50 percent.[37] In an ironic twist, when Mexicans seek employment in the same US industry that helped put them out of work, those lacking documents are rounded up, in some cases prosecuted, and deported. It's not just a metaphor—we have literally beggared those around us, and for some, validated the words of journalist Mignon McLaughlin: "Few of us could bear to have ourselves as neighbors."

But it's not just our close neighbors we hurt by dumping. It's almost any farmer in the third world. In virtually every developing country where local farmers eke out a living growing commodities that are subsidized in the United States, our policy contributes to reduced incomes, increased unemployment, loss of land, and a decline in quality of life.[38] In short, our subsidy programs damage third world economies. As one study found, "Subsidized agriculture in the *developed* world is one of the greatest obstacles to economic growth in the *developing* world."[39]

American officials recognize the problems inherent in dumping, even though restraining the subsidy leviathan may be beyond their power. As former US Deputy Secretary of Agriculture Charles Conner said of the 2008 farm bill:

> This farm bill just heads in the wrong direction in terms of
> our international obligations. It's no secret our current farm
> programs under current law have come under enormous fire
> for their adverse impact on developing regions of the world and
> their ability to increase their agricultural production because
> they can't compete against the farm subsidies of the developed

world. How does this bill respond? This bill responds by increasing trade-distorting supports on seventeen out of twenty-five of the commodities that we provide. This is moving, clearly, in the wrong direction in terms of helping the world sustain themselves through food production.[40]

What does the dumping of corn and soybeans have to do with meatonomics? Follow the money. Recall that in the United States, most of these crops are fed to livestock and fish—making American animal food producers the main beneficiaries of feed crop subsidies. These producers use their economic and political clout with remarkable success to convince state and federal lawmakers to create and maintain subsidies for feed crops. As noted, one study found that every $1 in campaign contributions that agricultural donors gave lawmakers yielded an investment return of about $2,000 in subsidy payments.[41] It's these subsidies, in turn, that cause the dumping of commodities used as livestock feed (like corn and soybeans) in the third world. As the United States is the planet's largest corn supplier, providing half the corn consumed by people and animals on Earth, our subsidy practices have huge consequences for the rest of the world.

Monsanto Appreciates Your Donation

We've seen that the massive handouts Americans give to big agriculture hurt taxpayers, consumers, and small farmers in both the United States and developing countries. The animal food industry is one of the rare beneficiaries of these subsidies, but it's not the only one. Those who provide key supplies to agriculture—known as inputs—also benefit from government subsidies that keep feed prices low. These input suppliers include equipment manufacturers like John Deere, chemical companies like Dow, and seed companies like Monsanto.[42] Crop subsidies encourage farmers to grow more crops, and of course the more they grow, the more seeds and tractors they need.[43] The fact that subsidies *seem* to be paid to the small farmers who consume these resources is merely an illusion; a political sleight of hand (remember the vastly undersized percentages that actually go to those farmers). Like a cruel practical joker, the government waves an envelope of cash in front of a small farmer—and maybe even lets him hold it briefly—but ultimately slips it into the pocket of an executive or shareholder in a large multinational corporation like Monsanto or Tyson Foods.

For their part, executives and shareholders in the handful of corporations that actually benefit from subsidies are living high off the hog. In the meat industry, shareholders in public companies enjoy nearly $4 billion in extra shareholder equity and roughly $228 million in annual dividends thanks to subsidies.[44] Tyson Foods, Inc., is the largest processor of chicken, beef, and pork in the United States and the second largest in the world, with more than a hundred thousand employees in more than 3,200 facilities. In 2010, Tyson paid shareholders $59 million in dividends; it also paid its CEO $4.8 million in compensation and its chief operating officer $4.9 million. Yet these and many other rich benefits were, in effect, paid completely by taxpayers. Research suggests that government crop subsidies in 2010 let Tyson pocket *at least* $3.5 billion in savings.[45]

The Bill We Love to Hate

Farm policy is like the weather: everyone complains about it, but no one does anything to change it. Every five years or so, Congress passes a massive farm bill with enough pork in it to satisfy just about everyone. In 1973, farm-state lawmakers shrewdly ensured the continuing support of urban lawmakers by adding food stamps to the farm bill. With more than 46 million Americans, or one in seven, now receiving food stamps, the farm bill is as firmly entrenched in American political life as apple pie and Social Security.

That is to say, entrenched but hated. As the *New York Times* noted following the 2008 farm bill's passage:

> Few pieces of legislation generate the level of public scorn consistently heaped upon the farm bill. Presidents and agriculture secretaries denounce it. Editorial boards rail against it. Good-government groups mock it. Global trading partners formally protest it. Even farmers gripe about it. But as Congress proved again last week, few pieces of major legislation also get such overwhelming bipartisan support.[46]

Republican Senator John McCain opposed the 2008 farm bill, saying: "It would be hard to find any single bill that better sums up why so many Americans in both parties are so disappointed in the conduct of their government, and at times so disgusted by it."[47] McCain reiterated his frustration four years later over the 2012 farm bill (which failed to pass Congress and was retooled as the 2013 farm bill), saying he was "hard-pressed to think of any other industry that operates with less risk at the expense of the American taxpayer."[48] Yet despite the occasional eleventh-hour drama, every farm bill eventually passes—mainly because there's a little something in it for everyone. It's a bizarre feature of US lawmaking that a bill's overall effect can be awful, but it will still pass if its individual components gratify enough special interests or regional constituencies. As Steve Forbes said, "Once a pork barrel scheme is started, nothing in heaven or on Earth is likely to stop it. Like barnacles on a ship, too many vested interests will glom onto it and fight to protect it."[49]

The market distortion and social disruption that farm handouts cause might be understandable if subsidies at least helped the taxpayers who write the checks. However, taxpayers are also major losers in the American farm subsidy system. According to one group of researchers, our subsidy program merely "shifts the burden of supporting farmers from . . . purchasers to taxpayers."[50] How much are these subsidies really worth to taxpayers, in their alter-ego roles as consumers, at the cash register? Not much, it turns out. An estimate in Appendix C shows that eliminating three-quarters of US animal food subsidies would raise prices by only about 1.8 cents per dollar of retail sales.[51] By contrast, for each dollar of animal foods sold at retail, producers shift $1.70 in external costs to taxpayers. In other words, taxpayers get almost none of the benefits of subsidies but all their costs—including, particularly, the damaging market consequences that subsidies foster.

Most commentators agree the agricultural subsidy system needs fixing, and many believe the best fix is to return to an unsubsidized market. But even if lawmakers suddenly found the political will to start dismantling the subsidy program, which is unlikely without massive public pressure, there would be no quick fixes. What does all this mean for American consumers? Many are challenged just to support their loves ones, or as George W. Bush clumsily said, "to put food on your family." Few want to see subsidies cut if the result is to raise meat and dairy prices. Yet there is a solution that, evidence shows, can benefit society across the board by lowering external costs, reducing disease, saving lives, and protecting the environment. That proposed solution is discussed in chapter 10. In the meantime, let's explore the other costs that the animal food industry imposes on society.

Food for Thought

- Over the past century, animal food producers have implemented improvements in efficiency that have boosted the productivity of animal farming and helped lower the retail prices of animal foods. However, these apparent gains are offset by a massive, growing pool of deferred, unpaid costs. Each year, producers of meat, fish, eggs, and dairy impose $414 billion in externalized costs on American consumers and taxpayers.

- As a category, farm subsidies represent one of the most significant costs that the animal food industry drops on Americans' shoulders. At $38.4 billion yearly, subsidies to animal agriculture allow a large animal food producer like Tyson Foods to save $3.5 billion or more each year in operating costs. Subsidies also generate more than $4 billion in extra shareholder equity for the small group of public companies that benefit from them.

- While farm subsidies provide a huge financial boost to animal agribusinesses and their suppliers, these corporate welfare programs disrupt markets and hurt almost everyone else they touch. Thus, subsidies foster overproduction, the rise of factory farms, and increases in externalized costs. As a result, subsidies hurt consumers, taxpayers, small farmers, rural Americans, and people in developing countries.

Diseases and Doctor Bills

The Land of the Free, by some measures, is becoming the Land of the Sick. One in three American adults has heart disease (including hypertension, or high blood pressure).[1] One in nine has diabetes, and one in twenty-five has cancer.[2] Over the past several decades, even as advances in medicine helped eradicate most infectious diseases in the United States, we've developed a new set of afflictions: diseases of indulgence. Americans spend more per person on health care than any other people on the planet, but we're far from the healthiest. And as we'll see below, a significant part of the nearly $1 trillion in annual health care and lost productivity costs related to just three diseases—cancer, diabetes, and heart disease—is directly linked to high consumption of animal foods.

The United States isn't the only country with health problems. In a report explaining why 1 billion people on the planet are overweight and prone to disease, the World Health Organization blames reduced physical activity and "foods with high levels of sugar and saturated fats."[3] The equation is simple: lack of exercise plus poor diet equals health problems.

But for Americans, our food choices are far from simple because of the producer-driven forces of meatonomics. As much as consumers would like to stay informed and eat what's right, we often don't get the proper incentives or information. This chapter explores the relationship between animal foods' low prices, Americans' heavy consumption of those foods, and the resulting diseases and financial costs. The idea that money is at the root of the problems in meatonomics is one of the central themes of this book. Yet, the real kicker is that financial drivers, perhaps as much as—or more than—lifestyle or dietary preference, are a key reason why Americans consume so much meat, fish, eggs, and dairy. As the Pew Commission noted in a 2008 report on factory farming, "Animal-derived food products are now inexpensive relative to disposable income, a major reason that Americans eat more of them on a per capita basis than anywhere else in the world."[4]

The artificially low prices of animal foods encourage Americans to consume these items at levels much higher than normal market forces would dictate. Government-led marketing programs add fuel to this fire of consumption. As a result, Americans eat an average of 0.6 pounds of meat each day, or about four USDA-measured servings. While this might sound like a healthy daily ration, it's actually more than a typical adult needs—by even the most liberal measure—and more than many can safely process.

In large part because of our high meat consumption, Americans regularly exceed USDA guidelines for intake of both protein and fat. Adult males under fifty, for example, eat twice the recommended daily protein allowance and close to twice the recommended maximum of saturated fat.[5] (And that's ignoring the fact that, as we'll see, the USDA's recommended levels are artificially high and represent poor targets for healthy consumption.) People don't *need* to consume meat at these levels. But market forces motivate us to do so, and rational consumers simply follow market cues. Unfortunately, those cues are making America ill.

Meat and Disease

Consider the case of the former evangelist of ultra-high meat consumption, Dr. Robert Atkins. In 2002, the Atkins Diet's founder and chief proponent had a heart attack. Rather than let the ailing physician recover in peace, critics seized the opportunity to speak out against the low-carb, high-fat diet he had followed for years. Atkins denied his diet was to blame, instead citing a chronic infection. But when bad luck visited the doctor again the following year and he died after a serious fall, the coroner's report noted that he had a history of heart attacks, congestive heart failure, and high blood pressure—all associated with eating too much saturated fat.[6] He was six feet tall and weighed 258 pounds at death, yielding a body mass index of 35 and placing him in the severely obese category. The Atkins Diet may not have single-handedly killed its founder and chief proponent, but it seems to have caused a number of life-threatening health problems likely to have killed him eventually.

This book is not meant to dispense nutritional advice. However, because I argue meatonomics is making Americans ill in record numbers, a few nutritional highlights (or lowlights) are necessary to show why that's the case. For starters, most of the research in meat consumption in the past several decades shows that the more of it people eat, the more likely they are to develop disease. This research is presented in hundreds of studies published in peer-reviewed journals like *The New England Journal of Medicine* and the *American Journal of Epidemiology.* (Google is a great tool—if you want to read any of the articles cited in this book's endnotes, they're easy to find.) Of course, you won't find this research discussed in the marketing communiqués issued by meat and dairy producers, and if you haven't heard it before, that may be why.

The point isn't that it's a miracle you're still alive if you've eaten meat for years. Rather, the depth and breadth of studies leave it beyond dispute that the levels at which Americans consume meat and dairy increase our incidence of disease. And this higher incidence of disease, in turn, generates economic costs that affect us all.

Take heart disease, the number-one killer of Americans, dispatching more people each year than AIDS, cancer, and car accidents *combined.* A stack of published studies taller than a quadruple cheeseburger with all the fixings establishes a direct, causal link between eating meat and developing heart disease.[7] Much of the research finds that red and processed meats carry the biggest risk of heart disease, making these the most dangerous animal foods. But a 2008 study published in the *European Journal of Clinical Nutrition* found those who eat poultry just twelve times a month are more than three times as likely to develop heart disease as those who rarely eat it.[8] A typical American eats 69 pounds of poultry per year, or about thirty servings per month.[9] Thus, this study shows that at *less than half* the amount the typical American eats each month, consumption of chicken and turkey can cause heart disease.

For those who've read that turkey burgers and boneless chicken breasts are healthier than steak and hamburger, this news may come as a shock. It turns out that *healthier* is a relative concept. Clinical studies comparing one animal food to another do show some are better than others, but that's like saying it's healthier to smoke filtered cigarettes instead of unfiltered. In fact, when studies compare the health effects of plant foods to animal foods, including fish, animal foods always lose.[10]

Consider dietary cholesterol, which elevates blood (or serum) cholesterol and causes heart disease.[11] Plants contain no cholesterol, but as shown in table 6.1, fish, poultry, and red meat contain a little dose in every forkful. In fact, bite for bite, according to the USDA, salmon and chicken contain about the *same amount* of dietary cholesterol as ground beef. Is it any surprise that cardiologists increasingly advise those with advanced heart disease to avoid *all* animal foods?

TABLE 6.1 Cholesterol Content of Select Animal Foods (mg of cholesterol per g of food)[12]

FOOD ITEM	CHOLESTEROL (MG/G)
Fish, pollock, walleye; cooked, dry heat	1.0
Chicken, broilers or fryers, thigh meat only; cooked, roasted	0.9
Beef, ground, 85% lean meat / 15% fat, patty; cooked, broiled	0.9
Beef, ground, 75% lean meat / 25% fat, patty; cooked, broiled	0.9
Fish, salmon, sockeye; cooked, dry heat	0.9
Turkey, all classes, meat only; cooked, roasted	0.8
Fish, haddock; cooked, dry heat	0.7
Fish, flatfish (flounder and sole species); cooked, dry heat	0.7

Another disease closely linked to animal food consumption is type 2 diabetes, the scourge of some 34 million Americans. A 2011 study by researchers from Harvard Medical School, Harvard School of Public Health, and other leading institutions looked at the eating habits of more than two hundred thousand people. The study, published in *The*

American Journal of Clinical Nutrition, concluded that consumption of just one daily serving of red or processed meat was associated with up to a 35 percent higher risk of type 2 diabetes.[13] Thus, if eaten as a steak, hamburger, hot dog, or bacon, a *single serving* of meat each day—well below the multiple servings Americans actually eat—significantly increases one's risk of becoming diabetic. As Americans eat about 50 percent more red meat than white, such research may help explain the nation's surging incidence of type 2 diabetes.

Or take cancer. Regularly eating the amount of meat in three Chicken McNuggets, about one-tenth of the typical American's daily meat intake, is enough to materially increase one's risk of developing cancer.[14] Copious research finds that meat-eaters are particularly prone to cancers of the prostate, breast, and colon.[15] Again, the studies link cancer more to red and processed meat than to white meat, and again, our nation's prodigious consumption of red meat may help explain our nation's high cancer rates.

It's not just a few outliers tapping on typewriters in the middle of a forest who have established these links between meat and disease. Four centuries ago, clergyman Thomas Fuller noted with surprising acuity: "Much meat, much malady." Since then, his view has found support in hundreds of clinical studies from around the globe. If a single daily serving of meat can increase one's risk of developing cancer, diabetes, or heart disease, the four servings each American actually *does* eat are just asking for trouble.

Incidentally, the list of diseases clinically linked to meat consumption goes well beyond just the big three of cancer, diabetes, and heart disease. Alzheimer's.[16] Parkinson's.[17] Crohn's.[18] Arthritis.[19] Gout.[20] Cataracts.[21] Don't take my word for this. Pick any article cited in the endnotes, type the title into a browser search field, and digest the news for yourself.

But what about the important nutrients people seek from meat, like protein and iron? And what about dairy's nutrients, like calcium? How about the omega-3 fatty acids people get from fish? Or the benefits of eggs? While this chapter's purpose isn't to address the nutritional pros and cons of these animal foods, an analysis can be found in Appendix A (spoiler alert: these foods aren't as necessary as you might assume).

How We Stack Up

To put Americans' half-pound of daily meat in perspective, consider how we compare to others. Per person, Americans eat nearly three times as much meat as any other culture or country on the planet.[22] We also have nearly triple the cancer rate and double the rates of obesity and diabetes as the rest of the world.[23] Of course, anyone can throw around statistics, and sometimes numbers can mislead. One might argue that because of Americans' comparative longevity, we're simply more likely than those in other countries to live long enough to develop diseases associated with old age. Still, for the nation with the world's highest gross domestic product and the sixth highest per capita income (based on purchasing power parity), factors typically associated with longer life spans, our life expectancy is surprisingly unimpressive. In worldwide longevity rankings, we're about fiftieth.[24] Most of the countries that beat us in longevity (including every nation in Europe) also have lower disease rates, meaning our high rates of diseases of indulgence stem from factors other than old age.

Another comparison offers even more clarity: how vegetarians stack up against meat-eaters. When researchers compare vegetarians to omnivores, they routinely find significant differences in health and longevity between the two groups. You don't need a mystic who can see the future for this science-based prediction—compared to a vegetarian, the typical meat-eater will die as much as ten years earlier and have twice the risk of diabetes and half again the risk of heart disease and colorectal cancer.[25] Omnivorous men have two and a half times the risk of prostate cancer as those who abstain from meat, and omnivorous women have twice the risk of breast cancer.[26] And of course, because these diseases can kill, omnivores' overall risk of death is two to three times higher than vegetarians'.[27]

But what about people who have eaten a half pound of meat every day for years without any health problems? Don't years of disease-free consumption mean something? Not really. Diseases of indulgence don't develop overnight the way infectious diseases do. The incidence of meat-triggered diseases increases with age, and many simply haven't reached the age when these illnesses develop and start to show symptoms.

Of course, these diseases have various contributing causes, including genetics and nondietary environmental issues like exposure to toxins. But in the case of the big three—cancer, diabetes, and heart disease—diet seems to be behind one-third or more cases.[28] It's certainly *possible* to eat a half pound of meat every day and never develop disease. It's also possible to smoke two packs of cigarettes a day and never get emphysema or lung cancer. But in both cases, just like trying to beat a casino at its own game, the odds are against it in the long run.

Obeying the Law of Demand

When the price of a product falls, we generally buy more of it. When the price rises, we buy less. That's the law of demand in a nutshell. Just how closely our behavior tracks a good's price depends on many factors, including our disposable income, how badly we want it, whether substitutes are available, and so on. Take cigarettes. One study found that a 10 percent increase in cigarette prices lowered the number of high school students smoking by 3.5 percent.[29] As simple as the law of demand is, it has profound implications for meat and dairy consumption. That's because industrial production methods and other producer-spurred phenomena of meatonomics keep prices artificially low, and low prices mean more people in line at the butcher counter.

Recall that the inflation-adjusted retail prices of animal foods have declined over the past century. Since 1935, ham is cheaper by 48 percent and steak by 20 percent.[30] As the law of demand predicts, the steady drop in prices causes an upswing in consumption. Since 1935, US per capita meat consumption (for all meat types) rose by an astonishing 95 percent. In the chicken category, while prices fell to one-quarter of their original level, per capita consumption jumped sixfold. Chart 6.1 shows the dramatic relationship between US chicken prices and consumption during this period. (Note that because this consumption increase is measured on a per-person basis, it is unrelated to growth in the US population.)

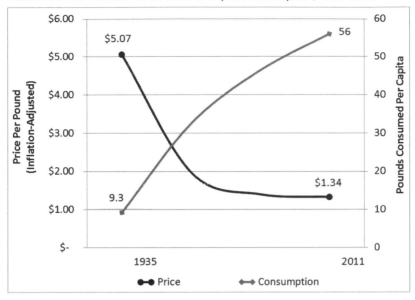

CHART 6.1 US Chicken Prices and Per Capita Consumption, 1935–2011

Of course, it's not fair to give *all* the credit for our high meat consumption to low prices. After all, people eat meat for reasons other than price—they might like the taste, or they might think it's good for them. Americans' disposable incomes have risen in the past century, which has helped spur consumption. And as we've seen, consumers are constantly hit with aggressive government messaging that urges us to eat more meat and dairy. However, while these non-price factors are important drivers of demand, price is also important. How important? Luckily, economists have a way to determine just *how* significant price is among the various factors that affect demand. It's relatively easy to measure the effect of price changes on quantity demanded in a manner that takes into account the effect of other factors.

The technical term is *price elasticity of demand,* and it describes how closely demand for a good follows its price. If a 1 percent change in price causes a 1 percent difference in quantity demanded, demand is said to be elastic (or technically, unit elastic). If a 1 percent price change causes less than a 1 percent demand change, as in the cigarette example above, demand is relatively inelastic. In other words, the more *elastic* the demand for a good, the more a price increase will cause quantity demanded of the good to drop.

A number of factors influence a particular good's price elasticity, including the availability of substitutes and the good's necessity (real or perceived). For example, because insulin is necessary for those with diabetes and there are no substitutes, demand for insulin might well be perfectly inelastic—that is, consumers might buy the same quantity regardless of price changes. The availability of substitutes is often a big factor in a good's price elasticity. As a general category, soft drinks have demand elasticity of about 0.8, which is relatively inelastic and means people want soda enough that price increases have less than a one-to-one effect in lowering sales.[31] However, because of widespread substitutes within the soft drink category, the Coca-Cola brand has price elasticity of 3.8.[32] This means that if Coke prices were to go up by 10 percent, consumers would switch to other brands or flavors and cause Coke sales to go down by a whopping 38 percent. Demand for Coke, then, is highly elastic—and brand loyalty of its drinkers, in the face of price hikes, is flatter than a week-old can of the stuff.

A hefty pile of studies (419 at last count) measures the price elasticity of demand for animal foods—that is, the relationship between prices and consumption. In 2010, a meta-study determined that a 1 percent change in the price of beef causes American consumption to change by about 0.75 percent.[33] In other words, notwithstanding all the other reasons why people buy beef—including marketing, taste, and nutritional beliefs—a 10 percent price change will shift consumption by about 7.5 percent. The study found that dairy has demand elasticity of 0.65, meaning that a 10 percent price change in dairy products causes consumption to change by about 6.5 percent.[34] Other animal foods are less elastic. A 10 percent rise in the price of eggs, for example, causes only a 2.7 percent drop in consumption. One reason why demand for eggs is less elastic than that for dairy is probably the perceived lack of egg substitutes—you might drink soy milk instead of cow's milk, but you're less likely to make a fried egg out of egg replacer.[35] (Although the idea that substitutes are lacking is really a problem of perception; tofu, for example, makes a mean breakfast scramble.)

The weighted-average elasticity rate for all animal foods is about 0.65, based on the aggregate wholesale, or production, values of beef, pork, poultry, fish, eggs, and dairy consumed in the United States yearly.[36] That is, on average, a 10 percent change in the retail price of an animal food causes a 6.5 percent consumption change. While this *relatively inelastic* figure means that price changes in animal foods have less than a one-to-one effect on quantity demanded, price is nevertheless still quite important as a driver of demand. As animal science professor Marta Rivera-Ferre notes, "consumer demand [for animal foods] is not linked with the actual biological needs of the human organism *but with prices.*"[37] With this elasticity figure in hand, we can make some curious observations about the effect of meatonomics on Americans' consumption of animal foods.

Retail Prices and Consumption Levels of Animal Foods

Economists are often interested in whether economic phenomena (such as a rise in consumption) are more supply driven (pushed by producer behavior) or demand driven (spurred by consumer income and preferences). In the case of animal food production, retail prices are low in large part because producers have gotten good at minimizing production costs through practices like hyper-confining animals and producing meat and dairy in high volumes to achieve economies of scale.[38] The elasticity data above suggest that retail prices are a large factor (although not the only factor) in Americans' decisions to buy animal foods. Since producers' cost-cutting behavior is a major driver in rock-bottom prices, it is also a major driver of consumer consumption. As professor Rivera-Ferre writes:

> It is wrong to believe that [animal food] production systems are . . . driven by consumer demands. It is probably more plausible the other way around, that the production systems are the ones that determin[e] and create the market. Thus, consumers adjust themselves to the offer, and not the other way around."[39]

Now, take this supply-driven influence on price and combine it with the routine use of misleading or confusing information by industry and government to sell animal foods (as seen in the book's first half). It's a double whammy. Consumers have faulty information, which makes fully informed choices nearly impossible, *and* we are seduced by artificially low prices. The combo leads to an almost preordained result: we consume much more animal foods than we would under more balanced and honest circumstances.

Using the elasticity figure above, we can estimate how consumption would be affected if retail prices reflected animal foods' true costs. We've seen that if these costs were internalized, requiring producers to pass all or most of them on to consumers rather than imposing them on society in roundabout ways, the true retail price of these products would soar to roughly $665 billion yearly from $251 billion. Thus, shifting all the externalized costs onto animal food producers would nearly triple animal foods' retail prices.

If a 1 percent price hike lowers consumption of animal foods by an average of 0.65 percent, simple math suggests a 170 percent price increase would reduce consumption by the maximum of 100 percent.[40] Would the entire country really stop buying meat and dairy if prices went up this much? No. The reason: some consumers *really* like steak and will pay almost anything for it. Nevertheless, it's clear that the act of shifting external costs to producers, with its ensuing price hikes, would lead to a sheer drop in consumption. Conversely, and more importantly, as long as the prices of animal foods stay *artificially low*, people will keep consuming much more of these goods than they would otherwise.

You can't blame consumers, because they're merely responding rationally to the cues that meatonomics provides. Under these cues, Americans consume much more animal foods than we would otherwise and considerably more than the government recommends. We've seen that Americans in practically every demographic group eat more from the USDA's meat group than they should—by as much as two-thirds above the USDA-recommended levels (see table 2.1 in chapter 2). Consider some of the side effects of this excessive consumption.

Much Meat, Much Malady

On a top-ten list of the worst health consequences of high meat and dairy consumption, excessive intake of cholesterol and saturated fat ranks high.[41] These substances are dangerous to our bodies in an insidious way, like creeping rot in a ceiling that makes the whole thing come crashing down one day. Right off the bat, it's important to note that neither cholesterol nor saturated fat is a nutrient, hence there is no USDA recommended minimum requirement for either. In a 1,357-page report commissioned by the USDA as the basis for its nutritional recommendations, the National Academies' Institute of Medicine notes that saturated fats "are *not essential* in the diet" and "there is *no evidence* for a biological requirement for dietary cholesterol."[42] Accordingly, the report says consumption of cholesterol and saturated fat should be "as low as possible."[43] That's because, among other things, "any intake greater than zero will increase serum levels of low density lipoprotein cholesterol, an established risk for cardiovascular disease."[44]

The federal government's recognition that *any* amount of ingested cholesterol or saturated fat is harmful has led to recommended daily maximums for these substances: 20 grams of saturated fat and 300 milligrams of cholesterol. However, although consumption in these amounts is already, by the National Academies' definition, harmful, Americans routinely exceed these levels. Males between forty and forty-nine, on average, routinely consume 35 percent more cholesterol than indicated. The USDA does not publish age-related cholesterol maximums, but the European Food Safety Authority (EFSA) does. Under EFSA guidelines, US teenagers ingest half again the recommended daily maximum of cholesterol for their age.[45] And as table 6.2 shows, saturated fat consumption is over the already unhealthy level of 20 grams in virtually every US demographic group.

I was once a poster child for an unhealthy relationship with fat and cholesterol—and my story, ultimately, illustrates the upside of breaking that vicious bond. For most of my adult life, like the typical adult American, my total cholesterol was at or over 200 mg/dl. I took a daily acid-reducing pill for years to combat chronic acid reflux, also known as Gastroesophageal reflux disease or GERD. In early 2008, I had a body mass index (BMI) of 25—the upper end of the healthy range. But after switching to a plant-based diet in the spring of 2008, my weight fell 10 percent, or about 17 pounds. My cholesterol dropped to 140. And with highly acidic animal products gone from my diet, the acid reflux ceased and I stopped the pills. These results might sound exceptional or apocryphal. They're not. They're normal, ordinary, and consistent with numerous clinical studies that compare omnivores and vegans.[46]

TABLE 6.2 US Daily Reference Value and Actual Consumption of Saturated Fat[47]

SEX/AGE	DAILY VALUE (G)	CONSUMED (G)	EXCESS CONSUMED
Male			
20–29	20	32.6	63%
30–39	20	33.6	68%
40–49	20	35.2	76%
50–59	20	32.4	62%
Female			
20–29	20	22.5	13%
30–39	20	23.6	18%
40–49	20	24.1	21%
50–59	20	22.9	15%

Fecal Contamination and Bacterial Disease

The predictable elements of cholesterol and saturated fats aren't the only things in meat that will do you harm. In fast-paced industrial slaughter operations, where workers process as many as three hundred animals an hour, a little feces invariably winds up in much of the end product. Such processing methods have given rise to a persistent epidemic of food-borne disease. This year, fecal pathogens will sicken one in four Americans and hospitalize three hundred thousand.[48] Two of the most common pathogens, campylobacter and salmonella, together will cause almost half of these incidents.[49] These organisms routinely live in the intestines of normal, healthy farm animals like pigs, chickens, and cattle, and are transmitted to humans via fecal matter. If you've ever had nausea, diarrhea, or a gurgling belly right after a meal, there's a good chance you ate something contaminated with live fecal bacteria. The problem has been eloquently summarized by Eric Schlosser, author of *Fast Food Nation*: "There's shit in the meat."[50]

In fact, the shit is hitting the fan. Fecal contamination is widespread, affecting all kinds of meat in virtually all US grocery stores, and the situation continues to get worse. A 2007 Consumer Reports study found campylobacter and salmonella bacteria in four out of five packaged chickens for sale in supermarkets and other retailers across the United States.[51] This discovery represented a big increase from the group's 2003 study, which found these bacteria in only half of chickens tested. Inexplicably, the expensive, organic chickens were more likely to contain salmonella than the inexpensive, inorganic chickens.[52]

Ground meat often contains fecal bacteria. One reason is that producers typically process multiple carcasses through a single grinder, giving bacterial contamination in the equipment an opportunity to spread from batch to batch. A USDA analysis of hamburger meat from hundreds of US slaughterhouses found fecal bacteria in four out of five samples.[53] In another study, federal agencies found *E. coli* in eight of ten ground turkey samples and seven of ten ground beef samples sold at hundreds of retail outlets throughout the United States.[54] Yet while ground meat is routinely contaminated, whole meat isn't much better. Thus, the feds also found *E. coli* in nearly nine out of ten chicken breasts and four out of ten pork chops sold.[55] Unfortunately, organically raised animals are usually butchered in factories at the same frantic pace as ordinary animals, so paying extra for the organic label doesn't make fecal contaminants any less pervasive.

By the way, don't be confused by the fact that fecal bacteria sometimes wind up on vegetables. In 2006, for example, more than two hundred people were sickened and three died from eating spinach contaminated with *E. coli*. But plants don't create feces; fecal bacteria begin their lives in the guts of animals. Thus, the 2006 spinach incident was traced back to cattle and pigs.[56]

Antibiotics

In factory farms across the United States, antibiotics are routinely fed to cattle, pigs, and poultry to promote growth and control disease.[57] In fact, in today's animal agriculture industry, where organic producers are as rare as hens' teeth, dosing animals with drugs is almost as common as feeding them. Unless and until the FDA takes steps to ban their use, these drugs will continue to figure prominently in Americans' meat, eggs, and dairy.

American farm animals consume 28 million pounds of antibiotics yearly, much of it to promote growth (rather than help the sick, as the drugs were intended). By comparison, Americans take about 7 million pounds of antibiotics per year, and then only to combat infection. For the consumer down the line, the practice of dosing farm animals leaves antibiotic residues in the meat, and it weakens the therapeutic effect of those antibiotics if needed someday to combat infection. Perhaps even worse, this practice of killing the weakest bacteria but letting the hardiest survive has given rise to drug-resistant bacteria that are difficult to combat in both animals and humans.

If the presence of residual drugs in your food makes you uncomfortable, the advocates at the Animal Health Institute (AHI), a trade group for animal-drug makers, are at the ready to reassure you. "Many published studies," states AHI in a promotional brochure, "have found that the risk to humans from resistant bacteria derived from eating meat or poultry from animals treated with antibiotics is *extremely minimal*."[58] Does this sound at all like the 1950s-era Chesterfield cigarette ads that informed readers of a "study" that found smoking Chesterfields did not "adversely affect" the "ears, nose, throat and accessory organs" of smokers?[59] In fact, a steady stream of published research starting in the 1970s shows that antibiotic use in animals causes a number of far-from-minimal health problems for humans, including allergic reactions, promotion of drug-resistant bacteria, and inefficacy of antibiotics when used therapeutically.[60]

The tendency for humans and farm animals to develop increased resistance to antimicrobial drugs is serious enough that in 1996, four federal agencies formed the National Antimicrobial Resistance Monitoring System (also known by the catchy acronym NARMS). Every year, NARMS analyzes drug-resistant pathogens in US retail meat and publishes its findings. Its latest report isn't pretty. There's been a proliferation over the past decade of various antimicrobial-resistant bacteria, including salmonella strains that develop in pigs, cattle, and chickens. These microorganisms have also grown steadily hardier and more drug-resistant since 1996.[61] If that's not enough to worry about, the report found an increase in unusually highly resistant *E. coli* strains in chicken breasts, ground beef, and pork chops as well.[62]

Beyond fostering new and more virulent strains of bacteria, the rampant combination of antibiotics and farm animals has promoted the innocuous-sounding but evil MRSA.[63] MRSA, or methicillin-resistant *Staphylococcus aureus,* is a staph infection that can lead to pneumonia, blood poisoning, toxic shock syndrome, heart-valve infection, and death. MRSA is particularly scary because of its resistance to methicillin, amoxicillin, penicillin, oxacillin, and other antibiotics. This highly adaptive organism responds to each new antibiotic it encounters by evolving genetically until the antibiotic is useless against it. Charles Darwin would be fascinated—this is natural selection, pure and simple. Single-cell bacteria evolve a lot faster than humans, and in case of MRSA, animal antibiotics are helping this organism win the race. Moreover, although we commonly associate the spread of MRSA with hospital wards, some of the disease's strains are known to have started among farm animals and spread to humans through contact with farm workers.[64]

A 2011 study found MRSA in nearly half of all meat samples collected from retail stores around the United States. Three-quarters of turkey samples contained the malevolent bacteria, as did two out of five samples of pork, chicken, and beef.[65] While cooking generally kills the bacteria and is advisable if you eat any industrially produced meat, those handling uncooked meat can become infected just from contact.

Hormones in Milk and Beef

The cocktail of farm animal drugs doesn't end there. Another chemical used to boost profits from industrial meat and dairy is growth hormone. As we saw in chapter 4, the FDA has approved the controversial growth hormone rBST for use in dairy cows. At the same time, most US beef cattle are implanted with a cocktail that may include sex hormones like androgens (testosterone substitutes), estrogens (primary female sex hormones), and progestins (agents that halt a female's estrus cycle). Some of these growth hormones are the same controversial and harmful anabolic steroids used by athletes to increase muscle mass. But forget looking like former home-run king Mark McGwire someday—when these dietary factors alter the human body's hormone balance even slightly, the effects can be damaging—particularly in fetuses and children.

One study found evidence of decreased fertility among children of mothers who ate large amounts of hormone-implanted beef during pregnancy.[66] Hormones in beef are also blamed for early puberty onset among children in Italy and Puerto Rico.[67] Because of such research, the European Union forbids all use of hormones in farm animals and even caused a mini–trade war with the United States by banning the sale or import of hormone-implanted beef. Yet the FDA, for its part, continues to permit hormone use in cattle, downplaying or ignoring the associated health risks.

We can blame meatonomics for the high levels of contaminants and residues in American animal foods. Humans managed to raise farm animals for millennia without giving them steroids and antibiotics, and as organic farming shows, we can readily do so when we choose. And when animals are slaughtered and processed carefully (although, as a result, at greater expense), feces are not routinely packaged with the meat. But meatonomics is a bulk business. It focuses on quantity over quality, on profitability over the health or comfort of humans and animals. Dosing animals with antibiotics and steroids helps the bottom line by increasing output, so it's developed into a bedrock meatonomics principle.

Adding It Up

We've seen the diseases. Now let's look at the doctors' bills. Based on research that shows the annual human medical costs related to *E. coli* and salmonella cases are $3.4 billion yearly, the portion of those costs associated with these pathogens transmitted by animal foods is an estimated $1.7 billion.[68] In the antibiotics department, the total annual health care cost related to antibiotic resistance in humans is roughly $47.2 billion, and the portion related to farm animals is estimated at half of that—or $23.6 billion.[69] Adding this number to the fecal contamination figures yields a ballpark total of $25.3 billion in annual, externalized human health care costs related to contaminants and antibiotics in our animal foods. But these costs, although huge, are just a drop in the health care bucket.

The annual costs of treating cancer, diabetes, and heart disease are massive—and growing. In 1980, total US health care costs were $706 billion (in today's dollars).[70] Today, we're spending three-quarters of that figure just treating the big three. Add in the additional expenses related to lost productivity—which we must do to gauge the full losses stemming from these illnesses—and the total costs of just these three illnesses rocket to more than $900 billion. Heart disease tops the list with total annual US costs of $477 billion,[71] followed by cancer with a total of $253 billion.[72] Type 2 diabetes is a close third at $184 billion.[73]

Of course, these diseases have a number of contributing causes, and diet is just one. Nevertheless, studies show that a diet high in unhealthy substances like cholesterol and saturated fats (that is, one based primarily on animal foods) is a significant factor in the development of these diseases. While definitive numbers showing the particular incidence of these diseases attributable to animal food consumption are not yet available, the medical literature does provide plenty of helpful guidance on the topic. Data support the estimates that a diet based on animal foods is responsible for roughly 30 percent of heart disease cases and roughly one-third of cancer and diabetes cases.[74] The cases of these three diseases caused by animal foods thus generate costs totaling roughly $289 billion yearly.[75] Adding to this figure the $25 billion in costs related to contaminants and additives brings the total to $314 billion. As the largest component of the external costs of animal foods, the US health care expenses of consuming these items could bankrupt a country. In fact, it's cheaper for India and Finland to run both their governments for a year.

CHART 6.2 Externalized Health Care Costs of US Animal Agriculture (in billions of dollars)

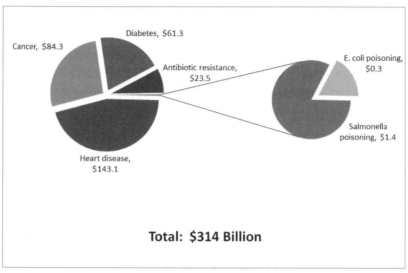

Total: $314 Billion

Is It Fair to Count These Costs?

It might feel like a stretch to assign medical costs to an industry because people who consume its products have health problems. If this concept troubles you, note that based on *actual* medical care costs, fifty US states have extracted a total of $400 billion from tobacco companies to reimburse state-paid medical costs related to smoking. That's a classic example of how an industry is sometimes required to bear some of its share of externalized costs.

Further, the $314 billion in costs of illness related to animal foods covers only a handful of diseases. Because reliable figures are not yet available for the costs associated with other diseases attributable at least in part to animal foods, like Alzheimer's, Parkinson's, and arthritis, this figure does not include any of those costs. But that doesn't mean these costs don't exist nor that they don't matter. As Einstein noted, "Not everything that counts can be counted." But it does mean that the number for these externalized costs is conservative.

Some may object that it's unfair to focus on the meat and dairy industries in assessing externalized medical costs. After all, every industry generates medical externalities, so why single out animal foods? Miners have high injury rates. Software programmers get carpal tunnel syndrome. Just as we do for the animal food industry, taxpayers and consumers help pay medical costs through taxes and insurance premiums for these other sectors.

But unlike these other examples, the animal food industry's hidden medical costs are unique and worthy of special attention. The external medical bills generated by meat and dairy are dramatically greater than those of any other industry. The hidden health care costs created by the next-highest sector, the tobacco industry, are $100 billion dollars *less* than those of animal foods. Moreover, the animal food industry's taxpayer subsidies of $38 billion make it one of the most heavily subsidized industries in the country.

Consider just how this diabolical dynamic works. Taxpayers provide the funding essential to keep most animal food producers profitable—and in some cases, to keep them in business. In turn, those same producers impose massive medical costs on us. In other words, first we pay factory farm operators to hurt us, and *then* we pay doctors to treat the injury. That blistering combination puts this industry in a special category all its own.

What can we do? The evidence shows that without question, Americans eat far too much meat and that this high consumption hurts our health and costs us money. One solution seems simple: eat less meat—or give it up altogether. This idea might once have seemed inconvenient, unpalatable, or extreme. But with the proliferation of plant-based foods and the rising number of high-profile vegans like former President Bill Clinton, business magnate Steve Wynn, real estate tycoon Mort Zuckerman, and Ford Chairman Bill Ford, that's no longer the case.

Food for Thought

- Animal foods' retail prices have fallen significantly in the past century (on an inflation-adjusted basis), driven largely by producers' cost-cutting practices. Compared to 1935, ham is cheaper by 48 percent and chicken by 74 percent. Just as basic economic theory predicts, these falling prices have helped boost consumption. Per capita chicken consumption, for example, has increased sixfold during this period.

- With these supply-driven forces spurring us to buy more animal foods than we would otherwise, Americans' extraordinary rate of animal food consumption is killing us in record numbers. We eat three times the meat per capita as others on the planet do, and as a result, we have nearly three times the cancer rate and twice the rates of obesity and diabetes.

- Diseases caused by animal food consumption generate massive financial costs related to treatment and lost productivity. At roughly $314 billion, the health care costs of animal food consumption refute the notion that the methods and levels of industrial animal food production are efficient. In fact, these costs tell of a bloated, overproducing industry that benefits from market failure and a resulting ability to sell its goods in artificially high quantities.

The Sustainability Challenge

The oldest river in North America, and the second oldest in the world after the Nile, is the ironically named New River. Cutting through the Appalachians like a winding road, the muddy, three-hundred-mile waterway meanders from North Carolina up through Virginia and West Virginia. Its sloping banks are lush with chestnut, white oak, sugar maple, currant, gooseberry, rhododendron, honeysuckle, and lily of the valley. Humans first settled at the water's edge more than ten thousand years ago, and as recently as the 1800s, native tribes including Shawnee and Cherokee fished and hunted in the area. But on a humid, early-summer day a couple of decades ago, as modern-day anglers waded the river's banks and recreational rafters screamed down stretches of whitewater, something happened that would change the New River for years.

On June 21, 1995, a factory farm's eight-acre pond of pig waste burst its retaining wall and dumped 25 million gallons of liquid manure into the New River. According to the *New York Times*, "Knee-deep red, soupy waste rushed over roads and tobacco and soybean fields and into two nearby tributaries of the New River until the waste lagoon, which is twelve feet deep and held waste from more than ten thousand hogs, was virtually empty."[1] It was the worst hazardous waste spill in North Carolina's history and one of the worst ever in the United States. After the spill, extraordinary levels of nitrogen and phosphorus in the New River caused persistent, toxic algal blooms and a massive hypoxic region—a dead zone in which oxygen levels were too low for the waterway's bass, trout, muskie, walleye, catfish, sunfish, and bluegill to breathe. The incident killed 10 million fish and closed 500 square miles of wetlands to shellfishing.[2]

While the manure spill at New River was unusually large, smaller-scale spills are surprisingly routine in the United States. A study published five years after the New River disaster counted a thousand manure spills in ten states over just a three-year period.[3] Animal waste regularly contains nitrogen and phosphorus, and of course, fecal bacteria like *E. coli* and salmonella. It often includes antibiotics and steroids. When these substances enter our water, as they do on a regular basis, they damage ecosystems and threaten the health of people and animals.

With 18,800 factory farms in the United States, routine water pollution is just one of a number of environmental side effects caused by Americans' insatiable demand for animal foods. Air pollution, land degradation, and climate change also figure prominently. Concerned commentators have proposed a variety of solutions to these problems, such as rotational farming, organic production, and local consumption. This chapter starts by exploring the viability and sustainability of these approaches. And because ecological problems impose hidden costs on taxpayers, the chapter's second half looks at factory farming's financial effects on the environment.

Not surprisingly, as with the other consequences of meatonomics, the factory farming practices that cause ecological damage are largely about the money. Critics "have been hard on hog farms," said Charles Carter Jr., whose company built the lagoon that broke and spilled into the New River. "People are just trying to make a living."[4] But when farmers and lawmakers say that livestock-related environmental harms are merely the inevitable result of trying to earn a living or feed the nation, they forget they are responsible for fostering the extraordinary and unnecessary level of consumer demand that drives these problems in the first place.

Farming and Nature

Occasionally, when I talk to someone about factory farming, I get an indignant rebuke. "I spent my childhood on a farm," the speaker admonishes me, "and it's not like that at all." She then goes on to describe the green, sunny pastures and happy, grazing animals of her youth. Sure, it was like that once. But those small, pastoral farms that some remember from the 1950s, '60s, or '70s, where animals slept in barns and spent their days outside, are almost all gone. They've mostly been turned into housing subdivisions or planted with monocrops like soy or corn to feed farm animals. Today, American farms are radically different.

Even before corporate players came to dominate American agriculture, farming was never especially natural or environmentally friendly. As plant geneticist Nina Fedoroff has noted, "agriculture is more devastating ecologically than anything else we could do except pouring concrete on the land."[5] That said, industrial agriculture has eliminated even the tenuous and limited balance that formerly existed in traditional animal farming. Livestock were once attached to land, where they ate plants and returned waste to the soil as fertilizer. But agribusiness executives discovered that decoupling animals from real estate allows greater scalability and productivity. With the link between land and animal lost, gone even is the modest symmetry formerly found in recycling manure and crop residues back into the system.

As practiced in the industrialized world, animal agriculture today is one of the least natural of all human endeavors. Factory farming now ranks with mining, oil production, and electricity generation as one of the most ecologically damaging industries on the planet.[6] In fact, in a much-quoted study, the UN Food and Agriculture Organization reports that among all industrial sectors, livestock production is "one of the top two or three most significant contributors to the most serious environmental problems, at every scale from local to global."[7]

Sustainability Solutions

With mountains of evidence showing that animal agriculture is polluting and warming the planet, well-meaning commentators have offered a number of ideas to shrink animal foods' ecological footprint. The recipe for sustainability in animal agriculture is variably said to favor small farms over big, organic over conventional, or local over remote. A look at some of these ideas helps answer the big question: can animal foods be produced sustainably to meet demand? (For purposes of this discussion, *sustainability* means the animal food system's capacity to endure without diminishing future generations' wealth, welfare, or utility derived from environmental resources.)

At the outset, note that measuring the environmental effects of human behavior is sometimes harder than you'd think. Take the conventional wisdom that says paper cups are more eco-friendly than Styrofoam. In fact, a paper cup requires thirty-six times more electricity and twelve times more water to produce than a Styrofoam one.[8] Styrofoam can take centuries to decompose, depending on the presence of moisture, sunlight, heat, and other factors. However, although not as persistent, a paper cup can still take decades to biodegrade because of coatings used to improve its ability to hold liquids. So which is better for the environment? "The truth" said Oscar Wilde, "is rarely pure and never simple."

"Sustainable" Meat Production?

Consider one of the best-known proposals to make meat production sustainable. In *The Omnivore's Dilemma,* Michael Pollan discusses ecological rotation as an alternative to factory farming. Pollan profiles Polyface Farm, a 550-acre organic farm in Virginia that raises animals by moving them around the farm on a regular basis in portable enclosures. The cattle are moved to fresh grass each day, and the chickens follow several days later to scatter the cow manure and eat the insects it nurtures. The farm produces 140,000 pounds of meat yearly, enough to feed seven hundred Americans at our average annual consumption level of 200 pounds per person.[9] Polyface Farm also supports the local food movement by refusing to ship meat beyond a four-hour driving radius. The farm's meat production methods are clearly more environmentally sustainable than most and, except for their practice of slitting the throats of alert chickens, generally more humane. And it's certainly an interesting novelty for a small cadre of well-intentioned consumers and restaurateurs in Washington, DC, and Richmond, Virginia, looking for local, organic meat.

But for a nation of more than 300 million meat-eaters, four out of five of whom live in cities, farms like Polyface are unlikely to ever become more than a novelty. For starters, Polyface produces only enough meat to feed a large college dormitory. Even if there were more than a handful of Polyface-like farms in the country, the quantity of food they could produce compared to factory farms would remain minuscule. The reality is that unless something is done to dramatically reduce the nation's current, extraordinary demand, the only alternative is to continue to industrially produce virtually all the meat, fish, eggs, and dairy that Americans consume.

Returning to Polyface Farm and other efforts like it, the big problem with the prominent attention they receive, both in *The Omnivore's Dilemma* and elsewhere, is that this focus exaggerates their relevance. Many think Polyface is representative of organic farming. It's not, by a long shot. Even if organic animal farming *were* particularly eco-friendly—which, as we'll see, it isn't—Polyface is far from an everyday example of the genre. That's because unlike the happy, grazing animals at Polyface, the vast majority of organic animals are raised in CAFOs largely indistinguishable from inorganic factory farms. For example, organic animals must—in theory—be given access to the outdoors. However, such access is largely meaningless both for organic pigs, who don't graze and are routinely fitted with nose rings to discourage rooting (by making it painful), and for organic chickens, who by nature don't dare venture outside the dark warehouses where they are raised. Pollan observed, for example, that during his visits to free-range chicken farms, he never actually saw a bird go outside.[10] In practice, the organic label signifies only the absence of synthetic pesticides in feed and chemicals in meat—it certainly doesn't mean the meat was sustainably or humanely produced.

Pondering Polyface

Even a close look at Polyface's ecological-rotation practice suggests that the system's reputation as a model of sustainability comes up short. For starters, Polyface is not self-sufficient, as all of its animals except cattle receive supplemental feed grown off-site. Adam Merberg, a UC

Berkeley doctoral candidate in mathematics, has calculated the calories in the supplemental grain fed to Polyface's chickens.[11] Merberg's numbers indicate that for all the Polyface eggs and chickens produced, more than three times as much food energy goes in as comes out. In other words, it's three steps back for every one step forward. That's not a great paradigm of sustainability. At 3:1, Polyface Farm's ratio of energy input to output for chickens is only slightly better than the 4:1 average in US chicken production.[12] As Americans increasingly shift from eating red meat to poultry, this is a major limitation in the Polyface model.[13]

Another problem with Polyface is its limited production capacity. For example, take the meat-eating demands of Southern California, population 23 million. Unless something is done to significantly lower their consumption, feeding these Californians the 14 million pounds of flesh they eat daily would require an additional thirty-three thousand farms the size of Polyface and an extra twenty-eight thousand square miles of farmland to contain them. That's more than the total area in seven of Southern California's eight counties.[14] This farmland simply does not exist in Southern California. Most of the region is surrounded by ocean or desert, except for the Central Valley to the north—which is already dedicated to providing one-eighth of the nation's agricultural output.[15] Even if part of the Central Valley could be converted to eco-rotation farms for meat production, most of that land is needed for crop production, and in any event, the entire area is just a fraction of the total that would be needed just to meet local demand. Further, if we were to try to feed the entire nation using the Polyface model, we'd need another 450,000 farms on an extra 390,000 square miles—an area almost twice the size of Texas. To quote Richard Oppenlander, author of the book *Comfortably Unaware,* the vast amounts of land needed for pasture farming make it "absurd" to think that the system can be sustainable.[16]

As much as we might hope Polyface Farm is the cure-all for the ills that stem from farming animals, it cannot serve as the model for animal agriculture. The small operation may work well in its local setting to serve a few hundred regional consumers, but the model isn't scalable to satisfy America's extraordinary consumption of meat—particularly in light of the country's mostly urban population. In fact, the mathematician Merberg reports that when he asked Polyface's founder Joel Salatin about the farm's lopsided input/output ratio during a live question-and-answer session in California, Salatin confirmed that Polyface is not sustainably self-sufficient.[17] Salatin, self-appointed evangelist of the rotational farming movement, is one of the most progressive, capable, and well-informed farmers on the planet. And if he can't find a way to make rotational farming self-sufficient and sustainable, it's unlikely anyone can.

Manic for Organic

Organic agriculture shuns manmade pesticides and fertilizers, and conventional wisdom says that makes it eco-friendly. That's one reason why organic foods represent the fastest-growing food category in the United States, with sales jumping from $1 billion to $26.7 billion over the past two decades.[18] But is organic food really as good for the environment as we'd like to think? Despite Prince Charles's claim that organic farming provides "major benefits for wildlife and the wider environment," a 2006 British government report found no evidence that the environmental impact of organic farming is better than that of conventional methods.[19]

In fact, because of large differences in land needs and growth characteristics between organic and inorganic animals, it's hard to draw conclusions about the environmental benefits of one production method over the other. As table 7.1 shows, considerably more land is required to produce organic animal foods than inorganic—in some cases more than double. This higher land use is associated with higher emissions of ammonia, phosphate equivalents, carbon dioxide equivalents, and other harmful substances. Further, denied growth-promoting antibiotics, organic animals grow more slowly—which leads to higher energy use for organic poultry and eggs. Thus, as table 7.2 shows, when the overall effects of organic and inorganic animal production are compared, the results are notably mixed.

TABLE 7.1 Land Use Needs of Organic and Inorganic Animal Food Production (in acres)[20]

ANIMAL FOOD	PRODUCTION QUANTITY	ORGANIC PRODUCTION LAND USE	INORGANIC PRODUCTION LAND USE	LAND USE NEEDS OF ORGANIC PRODUCTION EXCEED INORGANIC BY:
Mutton	1 Ton	7.7	3.4	126%
Eggs	20,000 Eggs	3.7	1.6	124%
Poultry	1 Ton	3.5	1.6	119%
Beef	1 Ton	10.4	5.7	83%
Pork	1 Ton	3.2	1.8	73%
Milk	10,000 Liters	4.9	2.9	66%

TABLE 7.2 Organic or Inorganic Production—Which Is Better for the Environment?[21]

ENVIRONMENTAL EFFECT	PIGS	POULTRY AND EGGS	CATTLE
Energy used	O	I	O
Land used	I	I	I
Pesticides used	O	O	O
Water used	N	N	O
Erosion and land degradation	N	N	I
Carbon dioxide equivalents emitted	O	I	I
Phosphate equivalents emitted	O	I	I
Sulfur dioxide equivalents emitted	O	I	I
Nitrates emitted	I	I	I
Ammonia emitted	O	I	I
Nitrous oxide emitted	I	I	I

Legend:
O = Organic is better (based on lower use or emission)
I = Inorganic is better (based on lower use or emission)
N = No significant difference

We can see that poultry and eggs are mostly more eco-friendly when raised inorganically, while it's generally more eco-friendly to raise pigs organically. As for cattle, factors like methane emissions and water use make the comparison more complicated.

Take methane. Besides figuring prominently in many a fart joke, it's a highly potent greenhouse gas (although in its natural state, it's actually odorless). A single pound of it has the same heat-trapping properties as 21 pounds of carbon dioxide.[22] Organic cattle must be grazed for part of their lives, which means that unlike feedlot cattle, they eat grass. However, cattle rely more on intestinal bacteria when digesting grass than grain, and this makes them more flatulent—and methane productive—when eating grass. The result is that grass fed, organic cattle generate four times the methane that grain-fed, inorganic cattle do.[23]

Then there are the water issues. On a planet where water is not only the origin of all life but also the key to its survival, animal agriculture siphons off a hugely disproportionate share of this increasingly scarce resource. It can be hard to picture the quantities of water involved, so consider a few examples. The 400 gallons of water needed to raise a single egg fill a family-sized hot tub. The 4,000 gallons required to produce one hamburger is more than the average native of the Congo uses in a year.[24] And the 3 million gallons used to raise a single, half-ton beef steer would comfortably float a battleship.[25]

Pound for pound, it takes one hundred times more water to produce animal protein than grain protein.[26] The ratio is a little less lopsided when comparing animal protein to other forms of plant protein, but it's still on the order of ten-to-one or higher. Thus, while producing 1 ounce of beef protein might take 9,000 gallons of water (depending on the production method), 1 ounce of soy or potato protein can be grown on as little as 400 or 700 gallons, respectively.[27] With these different water use characteristics in mind, let's consider the argument that organic beef and dairy production is eco-friendly because it uses less water than inorganic methods.

Organic cattle require 10 percent less water than inorganic but still need 2.7 million gallons each during their lives, enough to fill 130 residential swimming pools. In light of the orders-of-magnitude difference in water needed to raise plant and animal protein, does a

10 percent savings for organic cattle really matter? Looked at another way, if Fred litters ten times a day while Mary litters only nine times, is Mary's behavior really *good* for the environment? The value of such comparisons is dubious.

One in eight people on the planet lacks sufficient water.[28] But shortages aren't confined, as you might expect, to the developing world. In July 2012, according to the National Oceanic and Atmospheric Administration (NOAA), two-thirds of the contiguous United States was in drought.[29] These conditions caused massive damage to crops across the country and left some people wondering if the Dust Bowl had returned (or if it ever left). In the largest such designation ever, the USDA declared more than a thousand counties in twenty-six states natural disaster areas. The agency also rated the year's corn crop, much of which was lost to the drought, poor to very poor.[30]

In Texas, the leading livestock-producing state, the clash between animal agriculture and water conservation has reached a symbolic critical point. While 2012 was dry, the prior year was even drier, bringing the worst one-year drought on record to the state. As Texans diligently produced their main agricultural product, cattle, the NOAA determined in 2011 that the entire state was in extreme drought.[31] Does it really matter that organic cattle use 10 percent less water? The water used for the 14 million beef cattle that come out of Texas annually, 40 trillion gallons, could cover the entire state under a lake almost a foot deep. It isn't just the use of water to raise farm animals that causes recurring droughts in Texas and the rest of the country—increasingly, for example, scientists blame climate change for such conditions. Still, the odd juxtaposition of high resource use and extreme resource scarcity, particularly in heavy animal farming states like Texas, is food for thought.

These factors lead to one conclusion: we must treat as highly suspect the claim that organic animal agriculture is sustainable. Organic methods are an environmentally mixed bag—sometimes slightly better, sometimes a little worse, and often the same as inorganic. But since animal protein takes many times the energy, water, and land to produce as plant protein, any modest gains from raising animals organically are largely irrelevant.[32] Shocked that organic production isn't the silver bullet of sustainability? If so, you may also be surprised to learn that local foods—the subject of many an eco-friendly claim— also come up short.

Loco for Local

Sustainability, locavores insist, requires that we consume locally. But the data often suggest otherwise. Food's carbon footprint is measured using a technique called "life cycle assessment" (LCA), which examines the carbon impact of every step or component in a food item's production and consumption. LCA measures water use, harvesting methods, packaging materials, storage and preparation techniques, and other factors. But spoiling the local food movement's heavy emphasis on what it calls "food miles" is the fact that transportation averages only 11 percent of total carbon footprint and is thus a mere fraction of most edible items' LCA.[33] By contrast, the act of cooking food typically accounts for 25 percent of its carbon footprint, while production accounts for another 17 percent of the carbon footprint.[34] In other words, a modest efficiency or inefficiency in either production or cooking can easily outweigh transportation's entire effect.

The LCA data lead to some startling conclusions about food miles and the merits of local consumption. For example, one study found that it's more carbon friendly for the British to buy lamb from New Zealand than to buy locally.[35] Lamb production is much more energy efficient in New Zealand than in the UK, in part because British production relies on fossil fuels while New Zealand production uses 64 percent renewable fuels. Thus, British lamb production requires 45,859 megajoules (MJ) of energy per ton of meat, while New Zealand production takes only 8,588 MJ per ton. Even after adding in the 2,030 MJ of energy needed to ship the New Zealand meat to the UK, New Zealand is still the clear winner at only 10,618 MJ for both transport and production—less than one-quarter of the British production requirement. This difference in energy consumption means New Zealand also wins in CO_2 output related to lamb production—just 688 kg/ton compared to the UK's 2,849 kg/ton.[36]

In another example of Kiwi production efficiency, the same study found it's more carbon friendly for Brits to buy their powdered milk from New Zealand instead of locally. New Zealand dairy cows are generally pastured and eat grass, while British cows are mostly confined and eat forage feed like hay and nutritional supplements known as concentrates. The fuel inputs needed to produce the British cows' forage feed and concentrates lead to major efficiency differences in milk production between the two countries. Thus, it takes 48,368 MJ of energy to produce a ton of powdered milk in the UK, but only 22,912 MJ in New Zealand. Even adding the 2,030 MJ necessary to transport the Kiwi powdered milk to the UK, the total energy used for both production and transport of the New Zealand product is 24,942 MJ—about half that in the UK. Again, New Zealand's lower energy use means less CO_2 output: just 1,423 kg to produce and deliver a ton of powdered milk to the UK, versus the British emission of 2,921 kg of CO_2 to produce the same ton of product.[37]

As these examples show, placing too much emphasis on food's local origin can easily cause one to overlook LCA components that have a greater effect on the environment. Such results led the New Zealand study's authors to criticize the practice of equating food miles with carbon footprint—a practice they say "ignores the full energy and carbon emissions from production."[38] The moral here isn't that we should completely ignore food miles in measuring food's ecological impact; we just need to exercise more discretion in how much importance we give those miles. As Texas State University professor James McWilliams observes in his book *Just Food:*

> Sure, it feels righteously green to buy a shiny apple at the local farmers' market. But the savvy consumer must ask the inconvenient questions. If the environment is dry, how much water had to be used to grow that apple? If it's winter and the climate is cold, was the apple grown in an energy-hogging hothouse? Is the local fish I'm ordering being hunted to extinction? . . . Distance, in other words, is just a minor factor to consider. In overemphasizing food miles, we have missed important

opportunities to think more critically about the fuller complexities of food production.[39]

When we settle for options like Polyface, we take our eye off the ball. Seeming panaceas like local, organic, or eco-rotated food just can't overcome the biggest challenge facing American animal food production, and the one that threatens to defeat any attempt at sustainability: our extraordinary level of consumption. With annual per-person consumption at 200 pounds of meat and 620 pounds of dairy, for a national total of 250 billion pounds of animal products, we simply lack the resources, technology, and market incentives to raise these products sustainably. The machinery of industrial farming is bursting at the seams, spilling animal emissions and production by-products across all environmental media—air, water, and land. The result is that after paying for animal foods at the cash register, Americans incur another $37 billion each year in hidden ecological costs.

Adding Up the Costs

Nailing down precise numbers is tough when calculating environmental costs, because not everything in nature can be monetized. As ecologists Peter Miller and William Rees note, "Most economic analyses are money- and market-based and are thus thoroughly abstracted from nature."[40] Money just isn't the answer to every question. After all, how do we measure the economic cost of a species becoming extinct? What's the cash value of the disappointment you might feel if fecal bacteria closed your favorite lake? There's no controversy-free way to count the losses, but that doesn't mean they aren't significant.

Another challenge lies in the difficulty of determining causality when agriculture disturbs complex ecosystems. "Colony Collapse Disorder," the mysterious disappearance of millions of US honeybees, is a serious problem for US agriculture. That's because directly or indirectly, one-third of the food we eat depends on honeybee pollination—giving those pollination services an estimated value of $215 billion worldwide.[41] In 2008, there were just 2.4 million honeybee colonies in the United States, down from 5.9 million in 1945.[42] These massive colony losses have already raised honey costs and beehive rental costs, hurt some beekeepers' incomes, put others out of business, and threatened to disrupt the production of crops worth $15 billion.[43]

One theory for the bees' disappearance is that with vast amounts of US cropland now dedicated to monocrops like corn and soybeans, foraging bees cannot find sufficient nutritional or seasonal variety to meet their needs. Another is that the prevalent use of pesticides on crops is killing the little pollinators because, when exposed to toxins, bees become disoriented and die within twenty-four hours. As these likely causes stem directly from US consumption of meat and dairy, this consumption, at minimum, seems a contributing cause in the bees' disappearance. But just how much, we don't know yet.

Industrial agricultural practices disrupt ecosystems in many ways, including some we're just beginning to understand. In certain cases it may be premature to measure and allocate costs. But we've got to start somewhere. Peer-reviewed research permits an estimate of about $37 billion in environmental costs associated with producing animal foods, although gaps in the research, and items that current research can't measure, suggest the true figure may be significantly higher. While those costs may be elusive (at least at present), there are some amounts we *can* measure.

Meet Your Dirt

Did the Dust Bowl ever really end? Many of the same farming activities—like overgrazing and overplanting—that led to American farmers losing millions of tons of topsoil in the 1930s persist today. In fact, soil loss is one of the biggest problems farmers currently face, affecting nine out of ten acres of American cropland.[44] Erosion robs dirt of organic nutrients like nitrogen that help plants grow, and it leaves the remaining soil unable to absorb water at proper levels. As a result, eroded farmland can become less productive by 25 percent or more.[45] Shifting soils also damage nearby ecosystems, buildings, and infrastructure. Livestock production, according to the UN, is to blame for 55 percent of US erosion.[46] Applying this percentage to cost data from a study published in *Science* shows that the externalized costs of livestock-related erosion, including things like flood damage and siltation of reservoirs, total about $15.4 billion yearly.[47] (Note that this figure does not include erosion's significant *internalized* costs—those absorbed by producers—which are caused by land's lower productivity. Adding those would nearly triple the total.)

Farming and Warming

A 2009 study by World Bank scientists Robert Goodland and Jeff Anhang blames livestock production for an amazing *51 percent* of human-caused greenhouse gas emissions.[48] One reason the study's estimate is so high is the researchers went beyond previous assessments to count emissions that even the Kyoto Protocol doesn't measure, like farm animal respiration. This astonishing study might just cast the 2012 drought, which many believe was a manifestation of global warming, in a new light as largely the product of factory farming. As Jonathan Overpeck, professor of geosciences and atmospheric sciences at the University of Arizona said of 2012's long dry spell, "This is what global warming looks like at the regional or personal level."[49]

Perhaps even more astonishing is that if humanity wants to avert the worst effects of climate change, we can do so relatively cheaply. The Intergovernmental Panel on Climate Change (IPCC), a UN-formed group of thousands of scientists from 195 countries, estimates we can mitigate climate change by spending less than 0.12 percent of our gross domestic product (GDP) yearly.[50] In other words, about one-tenth of 1 percent of annual GDP—for a cumulative total by 2030 of less than 3 percent of one year's GDP. Using 2011 US GDP of $15.1 trillion, applying the 51 percent multiplier introduced by Goodland and Anhang, and adjusting for inflation, the yearly climate mitigation cost related to animal foods would be about $9.4 billion.[51]

Pesticides and Fertilizers

We might call this "trickle-off economics," not to be confused with the trickle-down theory the Reagan administration promoted. Pesticides contain poison. Fertilizers contain nutrients. Both cause harmful damage to ecosystems and drinking water when they trickle off into our lakes, streams, and groundwater. Pesticides, for example, are linked to bird losses, groundwater contamination, and toxin resistance in pests. Fertilizer in surface water, for its part, causes drops in real estate values, decreased recreational water use, and higher spending to save threatened species. Because more than half of US cropland is dedicated to growing livestock feed, the damage from pesticide and fertilizer runoff is closely linked to these crops. The annual social cost of trickle-off economics? About $7.5 billion.[52]

Feces and Fumes

A factory farm produces two things: meat and manure. The meat goes off-site to be slaughtered and packaged. The manure stays behind in lagoons where it slowly cooks, releasing ammonia, methane, carbon dioxide, and hydrogen sulfide into the air. Needless to say, it's best not to live downwind from a factory farm, in the danger area where noxious fumes drift like a long, smelly tail. A study by the Minnesota Pollution Control Agency found that hydrogen sulfide levels could violate state air standards as far as five miles downwind from a CAFO, and ammonia levels could violate standards a mile and a half downwind.[53] For those living in the noxious zone, a shift in wind direction can mean a daily change in health and attitude. (Talk about watching weather forecasts *very* carefully.) As one Illinois resident told a local newspaper, "I could be out in the garden, and . . . have to run for the house if the wind switches direction. . . . One night the smell was so bad, I said to my wife, 'I don't even know if it's safe to go to sleep.'"[54]

Life in this zone can also mean significantly depressed property values. Studies find that real estate values drop significantly with proximity to factory farms.[55] All told, the total annual reduction in US property values caused by CAFOs is about $2.5 billion.[56]

A Bunch of Manure

The 10 billion land animals raised yearly in the United States generate 1 trillion pounds of manure, enough toxic waste to fill New York's

Giants Stadium two hundred fifty times.[57] Because it has nowhere else to go, this manure sits—sometimes permanently—in lagoons around the country. From there, either through catastrophic breaks or persistent leaks, a little of it inevitably finds its way into our rivers, lakes, oceans, and groundwater. Like an overflowing toilet, this massive volume of poisonous, high-maintenance waste is literally overwhelming our capacity to handle it. The EPA, for example, has found that groundwater sources in one-third of US states are contaminated with animal waste.[58] The annual hidden cost of the animal waste problem, as measured by the expense to repair leaky lagoons and spread stored manure over cropland, is $2.4 billion.[59]

CHART 7.1 Annual Externalized Environmental Costs of US Animal Agriculture (in billions of dollars)

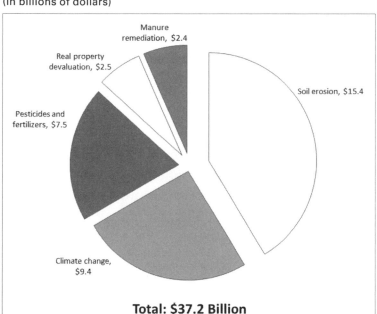

Manure remediation, $2.4

Real property devaluation, $2.5

Soil erosion, $15.4

Pesticides and fertilizers, $7.5

Climate change, $9.4

Total: $37.2 Billion

Chart 7.1 provides an overview of animal farming's yearly ecological costs in the United States. Compare these costs—$37 billion—with the $8 billion annual budget of the US Environmental Protection Agency, which is tasked with regulating CAFOs' environmental impacts. Does the EPA, which must also police dozens of other industries, have the necessary resources to monitor CAFOs? Perhaps more to the point, does the agency have the resolve to do so?

Where Were the Regulators?

It was mid-December 2008, just one month before Barack Obama was sworn in as the forty-fourth president. George W. Bush's appointee, Stephen L. Johnson, was still administrator of the EPA—but he didn't have long. Johnson's tenure had been controversial: among other things, he tried to block seventeen states from reducing greenhouse gas emissions and improving cars' fuel economy. This time, his agency was considering a hotly debated proposal that would exempt CAFOs from federal emissions reporting requirements. A few months earlier, the US Government Accountability Office (GAO) delivered a report to Congress with the lengthy but unambiguous title, "Concentrated Animal Feeding Operations: EPA Needs More Information and a Clearly Defined Strategy to Protect Air and Water Quality from Pollutants of Concern." The report found the EPA's efforts to collect data on CAFO emissions were inadequate, and as a result, the agency had failed to properly assess how "air and water pollution from CAFOs may be impairing human health and the environment."[60]

Nevertheless, on December 18, 2008, a lame-duck Johnson stunned environmental groups by adopting a rule exempting CAFOs from most reporting requirements for hazardous emissions. One immediate effect was that despite the EPA's own earlier finding that groundwater sources in one-third of US states were contaminated with animal waste, the agency began collecting *less data* on CAFO emissions instead of more as advised by the GAO. It's as if your kid was getting F's at school and you told the principal the best solution is to stop sending report cards home. Another result was that environmental groups brought a lawsuit challenging the rule. The court remanded the rule to the EPA for reconsideration and possible revision. Nevertheless, the eleventh-hour exemption adopted in 2008 remains in effect as of this writing.

It's a recurring theme in this book—regulatory policy heavily coated with industry-friendly flavor. The EPA is tasked with enforcing green laws like the Clean Air Act; the Clean Water Act; and the Comprehensive Environmental Response, Compensation, and Liability Act (CERCLA or Superfund). While the agency is technically responsible for requiring CAFOs to follow these and other laws, the EPA's rule-making practices have yielded enough industry-friendly loopholes to shock even some federal lawmakers.

Mull over another example from a few years earlier. In 2003, the GAO responded to an inquiry from US Senator Tom Harkin (D-IA) regarding CAFO-related water pollution. Harkin's concern might have stemmed from the 2002 study published in the *Journal of the American Water Resources Association* estimating that at least half of his home state's 729 regulated manure lagoons (and perhaps the same portion of its five thousand unregulated lagoons) routinely leaked more than the legal limit.[61] Or maybe what troubled him was the 2002 study by the US Geological Survey that detected animal antibiotics in four out of five waterways sampled near CAFOs in thirty states, including Iowa.[62]

In its response to Harkin, the GAO noted two fatal deficiencies in the EPA's CAFO program. First, because the agency's rules provide extensive exemptions, only two out of five CAFOs in the country are subject to EPA oversight.[63] Second, the agency has delegated its permitting functions to states but is ineffective at overseeing the states' activity. As a result, half the states audited by the GAO do not follow proper permitting procedures, and three states allow CAFOs to operate without permits—"leaving these facilities and their wastes," in the words of the GAO, "essentially unregulated by the [EPA's] CAFO program."[64]

The Case against Industrial Vegetables

It might seem unfair to single out meat and dairy producers for the microscope of ecological analysis. The reality is most modern farming methods are hard on the environment, regardless whether the product is meat, crops to feed animals, or fruits and vegetables for human consumption. Virtually all agriculture destroys topsoil and adds pesticides and fertilizers to the environment. However, we must keep in mind that, as the UN notes, "more than half of the world's crops are used to feed animals, not people."[65]

In the United States, the top three crops are corn, soybeans, and hay. Farm animals eat 70 percent of the soybeans, 80 percent of the corn, and virtually all of the hay.[66] Consequently, since our nation's farming effort is mainly focused on raising animals or crops to feed them, problems like land degradation and pesticide pollution arise mostly because of our insatiable demand for meat, not vegetables. As James McWilliams observes, "Every environmental problem related to contemporary agriculture . . . ends up having its deepest roots in meat production."[67]

Sure, growing vegetables hurts the environment too—but not nearly as much. That's because it takes five times as much land to feed an omnivore as an herbivore.[68] In Brazil, the world's leading beef exporter, the demand for land to raise beef cattle causes systematic destruction of the world's largest rain forest on an enormous scale. In 2010, the year with the *lowest* level of rainforest destruction in decades, Brazil clear-cut 1.6 million acres of rain forest.[69] That's a swath of jungle twice the size of Yosemite National Park. The economic forces driving this destruction are predominantly beef related: two-thirds of cleared rainforest is used for purposes related to cattle ranching.[70] To put it in basic math, two of the three acres of Brazilian rainforest cleared each minute are used to produce meat.

It takes more land to produce meat than vegetables because of the large volume of plant matter needed to raise animals. According to Cornell ecology professor emeritus David Pimentel, livestock must be fed 6 pounds of plant protein to produce 1 pound of animal protein.[71] Considering the various benefits of plants over meat, this strange alchemy is a little like turning 6 pounds of gold into 1 pound of lead.

Moreover, this lopsided input-output ratio might be exacerbating world hunger. Pimentel has estimated that if grain fed to livestock in the United States alone were instead fed to people, 840 million additional people could be fed—that's nine out of ten of the planet's undernourished.[72] An American eating a hamburger, in other words, uses enough grain to feed six hungry people. And paradoxically, while one in seven people on the planet routinely goes hungry, Americans eat animal foods at record levels—ensuring that two-thirds of our own population stays overweight.

Americans eat three times the meat per person as the rest of the world, but unfortunately for the rest of the world, they're catching up. As that trend continues, one day the world will be massively short of land to produce enough meat to satisfy global demand. Because of the tremendous amounts of real estate needed to grow feed for livestock, we'd need two-thirds more land than exists on the planet to feed the world meat at the extraordinary pace at which Americans consume it.[73] Accordingly, while industrial agriculture will always threaten the environment, it's clear that the main problem lies in the production of meat—not vegetables.

What It All Means

America's high production of animal foods is causing a head-on collision between demand for these items and the reality of scarce resources like land, water, and fossil fuels. This isn't a future threat; it's happening in real time, right now. The evidence is everywhere we turn, in every environmental medium. The most popular ideas proposed to address the problem—like eco-rotation, organic production, and local consumption—simply don't work. As discussed, it takes up to one hundred times more water, eleven times more fossil fuels, and five times more land to produce animal protein than equal amounts of plant protein. In short, the meatonomic system is not sustainable.

The EPA reports that each week, more than thirty square miles of productive US farmland are turned into homes or office buildings and lost to farming for good.[74] Because this regular loss of farmland makes it harder—and costlier—for producers to raise animals, they must devise ever more efficient production methods and locations. These include such environmental undesirables as clear-cutting forest and increasing factory farm output. Meanwhile, nearly 1 billion people remain hungry but might be fed if the 6 pounds of plant protein fed to animals for each pound of animal protein were redirected. Something's got to give. "To be perfectly blunt," writes McWilliams in *Just Food*, "if the world keeps eating meat at current rates, there's simply no way to achieve truly sustainable food production."[75]

The world's population is expected to grow from 7 to 10 billion in the next forty years.[76] The UN warned recently that worldwide production of animal foods is damaging the planet and, as human population and consumption levels continue to rise, will do even more harm in the future. The solution, according to the UN, is "a substantial worldwide diet change, away from animal products."[77] The menu is straightforward: less meat, fish, eggs, and dairy. The UN, by the way, is only the latest in a long series of commentators to advocate for vegetarianism to save the environment.

What can one person do? As one step, you might try consuming less milk and cheese. That's a logical option for the ecologically minded since, as we've seen, the nation's dairy farms are responsible for a large share of harmful methane emissions. And with animals generating one-fifth of the methane in the United States, the farting cow is no laughing matter.[78]

Food for Thought

- Animal agriculture drives massive, routine environmental damage that costs Americans more than $37 billion yearly. This sector's ecological impact is on par with the most destructive of industries, including mining and electricity generation, and its role in causing global warming exceeds that of even transportation or oil production. The externalized expenses of animal food production include costs related to soil erosion ($15.4 billion), climate change ($9.4 billion), pesticide and fertilizer damage ($7.5 billion), real estate devaluation ($2.5 billion), and manure remediation ($2.4 billion).

- A number of solutions are proposed to the environmental problems caused by animal production, but invariably, they come up short. Local foodists urge us to buy products raised nearby, but the data show that the carbon effects of transportation are often the least of our worries. Organic advocates say animals should be raised without chemicals, but in fact, organic production has as many environmental problems as benefits. And while eco-rotation evangelists argue that animals can be cycled around a farm in a sustainable way, the evidence shows that such systems require feed inputs at unsustainable levels and are not scalable to meet Americans' extraordinary demand. And as a reminder of the regulatory-capture theme that dominates meatonomics, industry influence at the US EPA has significantly diminished that agency's ability to control the damage caused by animal farming.

- Americans respond to the producer-driven forces of meatonomics by consuming 200 pounds of meat and 620 pounds of dairy per person each year, levels requiring that virtually all of our animal foods come from factories. At this pace, the environmental damage caused by animal agriculture can only worsen. Thus, any solution to this chronic environmental problem must begin by dramatically reducing our consumption of these items.

The Costs of Cruelty

Do fish feel pain? The issue has been hotly debated for years. One naysayer is former University of Wyoming professor James Rose. "A fish doesn't appear to have the neurological capacity to experience the unpleasant psychological aspect of pain," Rose wrote in 2000. "Thus, the struggles of a fish don't signify suffering."[1] Whether it's fish, cows, or chickens, this perspective that animals used for food don't suffer—at least not enough to worry about—is at the root of most modern animal farming practices. Thus, one North Carolina pig farmer said the hyper-confined pigs in his factory farm "love it. . . . They don't mind at all."[2]

Increasingly, however, consumers sense something is wrong with this dismissive attitude. People are informed and concerned about animal farming methods, and in surveys assessing shoppers' attitudes toward factory farming, a majority of respondents prefer practices that are more humane.[3] Beyond the weighty ethical questions, thanks to recent research, we can also now quantify and monetize consumer sentiment toward animal farming. For example, agricultural economists F. Bailey Norwood and Jayson Lusk show in their 2011 book, *Compassion by the Pound*, that consumers are actually willing to spend their own, real money—in average amounts ranging from $23 to $57 per thousand animals—to improve farm animals' lives.

Factory farming often exacts a toll on animals in the form of pain and stress, and because humans care about how animals are treated, we suffer too. In this chapter (and the closely linked Appendix D), you'll find recent research into the physiological and emotional effects that industrial production methods have on animals used for food, including fish, pigs, chickens, and dairy cows. You'll also find an estimate of the financial costs that producers impose on society by persisting in those methods despite a general consumer preference for change. A key caveat: I don't believe a simple cost estimate can meaningfully represent the massive scope of routine, lifelong suffering endured today by virtually all US farm animals (that is, the 99 percent of farm animals raised in factory farms). Nevertheless, because economists *do* recognize a human cost in known animal suffering, and this book is about economics, it's helpful to try to quantify this sum in an objective way.

To repeat a familiar refrain, *it's about the money.* Animal food producers don't think of themselves as a cruel or sadistic bunch. They're just trying to maximize profits, and in the animal food business, you do that by minimizing the amount spent on each animal's comfort. This turns the focus to animals as production units, rather than as living beings with physiological and emotional needs. Practices some consider cruel are just a side effect of industrial production. Ron Torell, for example, is a self-proclaimed "long-standing educator and advocate of animal agriculture" who writes a column for the Nevada Cattlemen's Association monthly newsletter. In a 2011 column, Torell advises beef producers to slaughter "poor producing economic units"—that is, low-yielding females—as soon as possible. Torell further cautions against letting "pet cows avoid the terminal trip to McDonald's. It makes no economic sense why these cows are given a free pass based on sentiment, color pattern or simply an experience the owners had with the pet when it was a calf."[4] Such a dollars-and-cents mentality informs most animal handling practices in the industry.

Some may question the appropriateness of this topic. For starters, cruelty is a value-laden, subjective concept, and what's cruel to one person or animal may not be to another. Some think it cruel to keep cats indoors while their natural urge is to be outside in the sunshine, exploring and playing. Others think it cruel to let cats outside, where they might kill birds or rodents, become lunch for a coyote, or get hit by a car. (I live with two cats and wrestle with this issue myself.) To avoid such value judgments, as used in this chapter, *cruel* means those practices which published research finds cause animals measurable pain or stress.

Some may find descriptions of cruelty disturbing. But because farm animal cruelty costs American society at least $21 billion yearly in externalized empathy costs, it's important to understand where these costs come from. As recent research on marine life is particularly timely and important, several pages of this chapter discuss fishing and fish farming. On the other hand, there is less material on decades-old practices like raising hens and pigs in battery cages and gestation crates, as most readers are likely familiar with them. (That said, for the uninitiated, a detailed summary of these methods can be found in Appendix D.)

The Cartesian Method

René Descartes, the 17th-century philosopher and mathematician, was one of the first scientists to theorize that animals don't suffer. The father of analytical geometry argued that among animals, humans were uniquely endowed with minds and souls. This belief led to his famous remark, *"Cogito ergo sum"* ("I think, therefore I am"). As a corollary, he argued that animals were mere machines whose responses to stimuli were simply mechanical. Because he and his followers considered nonhumans incapable of thought, emotion, or the capacity to feel pain, they treated live animals as pieces of equipment and mocked those who objected.[5]

However, today we know that if an animal changes her behavior in response to harmful stimuli, then she likely experiences those stimuli as pain.[6] Because the ability to feel pain helps animals remember to avoid harmful situations, this capacity has played a role in evolution and in the success of species who experience pain.[7] Conversely, if harmful stimuli did not produce a painful reaction, animals would not learn from encounters with such stimuli and would repeat dangerous interactions at the risk of injury or death. That's why it turns out, for example, that virtually all sentient beings have evolved the same ability that a human child has to avoid a hot radiator after touching it once or twice.

Fish and Pain

While the Cartesian view remains popular with some people in the scientific community, modern research has turned the traditional wisdom about the suffering of animals—including fish—on its old-timey wigged, powdered head. Scientists have confirmed in the past decade that fish not only feel pain, but they also experience emotions. In one study, researchers assessed how fish responded to having acid or bee venom injected into their lips. With eighteen pain receptors on a trout's head and face, some more sensitive than those in the human eye, it turns out they don't like having their face injected with stinging chemicals any more than we would. The injected fish engaged in stress-associated rocking behavior and, compared to control groups, reduced their swimming activity, waited three times longer to eat, and had significantly elevated breathing rates. The researchers concluded that "both the behavior and physiology of the rainbow trout are adversely affected by stimuli known to be painful to humans. This fulfills the criteria for animal pain."[8]

How do we know the trout exhibiting these responsive behaviors weren't just engaging in the reflexive behavior of Cartesian machines? Because in a similar, later study, trout dosed with morphine again had their lips injected with harmful toxins. The medicated fish engaged in significantly less rocking, lip-rubbing, and elevated breathing than those who did not receive morphine.[9] Researchers concluded that because the responsive behaviors were directly related to the level of pain as managed by the morphine, the trout's responsive behaviors could not be merely reflexive and that "pain perception in fish" is a reality.[10] Although these trout studies don't seem particularly humane in themselves, their results may suggest a need to reconsider how we treat marine animals.

Fish and Emotion

Another study involving goldfish shows fish experience emotions like fear and anxiety.[11] Researchers studied two groups of goldfish—one dosed with morphine and one undosed—who were subjected to painful levels of heat. Both groups responded reflexively in an attempt to escape the heat. However, after being returned to their tanks, the morphine-dosed group soon returned to normal behaviors, while the undosed group showed stress-related behavioral changes. The undosed fish acted defensively, exhibiting what researcher Joseph

Garner called "fear and anxiety." According to Garner, "The goldfish that did not get morphine experienced this painful, stressful event. Then two hours later, they turned that pain into fear like we do. To me, it sounds an awful lot like how we experience pain."[12]

Down on the Fish Farm

The recent research into fish and pain leads some to conclude that fish farming, the fastest-growing segment of animal agriculture, is one of the least humane of all processes to produce animal food. Farmed fish suffer routinely both during their lives *and* when slaughtered. As one would expect from any profit-minded fish farmer, tight stocking densities are used in typical farms to help keep costs down. But it's a hard-knock life for the fish since tight densities cause them chronic stress and make it impossible to engage in natural behaviors like defending territory or escaping from bullies. One group of researchers found that "the aquaculture environment is inherently unsuitable for fish that are territorial or solitary animals in their natural environment, such as some salmonid fish [salmon and trout]. In these cases, agonistic interactions can be particularly stressful to the fish."[13]

Packed stocking densities also cause fish a variety of physical problems. Injuries to tails and fins are common because of aggression-induced cannibalism and frequent friction with cages and other fish. Tightly packed fish are highly susceptible to eye diseases leading to cataracts and blindness, a problem pervasive enough to merit the formation of a group called Friends of Blind Fishes.[14] One research team even worried that the prevalence of blindness among farmed fish might give consumers "doubts on the ethical standards of industrial fish farming."[15]

These overcrowded containers can also give rise to parasitic infestations. In the case of salmon, there are various techniques for dealing with sea lice—none of which is completely effective and all of which have their own welfare implications. One is to douse infested salmon with a chemical like hydrogen peroxide. Because such chemicals are harsh skin irritants, they cause the fish to exhibit stress behaviors for days after treatment.[16] Another technique is to introduce helper fish called wrasse as cleaners to pluck the parasites from their hosts. However, in such close confinement, the wrasse are often bullied or killed by the salmon, and in any event, they're killed by farm workers at the end of the season to prevent the spread of disease to the next batch of salmon.

Slowly consuming their hosts, sea lice cause lesions, bleeding, and sometimes death. As one would expect, salmon don't enjoy being eaten alive; research shows those infected with sea lice suffer from chronic stress and compromised immune systems.[17] Sometimes the parasites eat all the way down to bone. When this happens on a fish's head, farm workers call the grisly result a "death crown."

Fish Kill, the Farm Way

When ready for slaughter, farmed fish are killed in profit-focused ways that many commentators deem inhumane. For starters, farmed fish are often starved for a week or more before slaughter to eliminate fecal matter from their intestines and make it easier to butcher them. While any sentient being presumably dislikes being starved, for fish conditioned to being fed at the same time every day, research finds this sudden disruption in their feeding schedule is particularly stressful.[18]

One common method of killing fish is to throw them in water rich in carbon dioxide. Placed into this acidic, low-oxygen environment, fish thrash around for half a minute or more, and even after calming down, continue to show signs of distress, such as vigorous head and tail shaking, for up to nine minutes.[19] Fish killed in this manner routinely bleed from the gills because of the intensity of their reaction.[20]

Another popular slaughter technique at fish farms is to bleed the animals while they're fully conscious. This might involve cutting open their gills, opening their bellies with a knife, or some other method developed for a particular species. It's unclear whether we need research to determine that it hurts animals to have their bodies cut open while fully conscious, but regardless, the research has been done. Here's what the scientists found: if not stunned first, fish feel pain when bled to death.[21] In fact, those eviscerated or degilled while conscious struggle "intensely" for four to seven minutes and respond to stimuli for up to fifteen minutes.[22]

For greater freshness and salability, many farmers like their fish to freeze while dying. Gradually slowing a dying animal's metabolism helps to minimize tissue decomposition and preserve its taste. Because fish asphyxiate at a slower rate when ambient temperatures are lower, chilling can lengthen the suffocation process by seven minutes or more.[23] Not surprisingly, being frozen to death is distressful to the animals. Research measuring levels of the stress hormone cortisol in fish found that these levels increase markedly when the animals' ambient water is chilled.[24]

Why Consider Fish?

In the world of animal foods, fish are an anomaly—an outlier. For one thing, fish differ from land animals in their inability to cry out when hurt or suffering. This powerlessness to vocalize leads many to confuse a dying fish's silence with a lack of suffering—although the research shows otherwise. And then there is the conventional wisdom that says fish are particularly nutritious. But a flotilla of scientific papers shows that fish are routinely high in mercury, PCBs, and cholesterol, making them a distinctly unhealthy alternative to plant foods (*see* Appendix A).

Nutritional issues aside, this chapter centers on the humane issues facing fish because marine animals frequently take a backseat to land animals—and because the recent research in this area is particularly compelling. Of course, cattle, pigs, and poultry have their own set of humane problems, like gestation crates for pigs, battery cages for laying hens, zero-grazing systems for dairy cows, and rapid growth and hyper-confinement for broiler chickens (as mentioned, more on that in Appendix D).

Measuring Cruelty's Costs to People

Given the chance, what—if anything—would you pay to change animal food production practices that are particularly inhumane? Economists Norwood and Lusk have sought to answer this question through studies involving real people and real money, and the answers are enlightening. In auctions where participants used their own cash to bid on animal welfare improvements, people paid an average of $57 per person to actually move one thousand laying hens from caged to free-range systems.[25] Bidders also spent an average of $23 per person to actually move one thousand sows and their offspring to shelter-pasture systems from confinement crates. Even more interesting for our purposes, Norwood and Lusk extrapolated from their auction results to estimate that people *would* spend a one-time average of $342 and $345 per person, respectively, to implement these two welfare changes throughout the United States.[26]

Now I propose to extrapolate further. Let's add three more hypothetical changes in animal farming to the two above: ending zero grazing for dairy cows, eliminating rapid growth and hyper-confinement for broiler chickens, and banning overstocking and inhumane slaughter of farmed fish. In terms of the number of animals affected, these five items likely represent the most prevalent industry practices in need of reform. Take $343.31, the midpoint of the range between the two amounts Norwood and Lusk estimated people would pay to improve hen's and pig's lives, and apply it to all the hypothetical changes. The total that this exercise suggests each American would be willing to pay, on average, to make these five changes is $1,717.[27] Adjusting this figure for inflation, multiplying by the number of US adults, then amortizing the total over twenty years (the standard IRS depreciation period for farm buildings) yields a total of roughly $20.7 billion yearly that farm animal cruelty imposes on Americans in externalized costs.[28]

Some will argue this figure is too high because not everyone would pay nearly $2,000 to improve the lives of fish and farm animals. That's true, but this figure is proposed as an average that puts us in the right vicinity. Some would spend nothing, while at the other extreme, some would spend $50,000 or $100,000. How much might billionaire casino owner and vegan Steve Wynn pay? Five million dollars? Fifty million?

In fact, if anything, I believe this cruelty number is too low. For starters, it excludes the amounts the animals themselves would be willing to pay—if this could be measured and conventional economics ascribed any value to it. Measuring animals' economic preferences is not all that far-fetched. One study actually measured pigs' willingness to pay for certain items. The animals were taught to repeatedly press a nose-plate to receive either food or increased social contact. By a ratio of 2:1, the pigs demonstrated they were more willing to spend time and effort on food than on friendship.[29]

Of course, if we could assign a value to animals' willingness to pay for better conditions, it would nevertheless go unrecognized by economic standards that measure only the *human* value of goods and services. Some believe this omission represents a deficiency in conventional economics. As methods improve for measuring animals' willingness to pay, and humans become more comfortable with the idea of using such figures, better metrics may emerge for making these calculations. Is it possible to measure the economic effects of producing animal foods without accounting for a single penny of economic cost associated with the individual animals' personal suffering? Because many of the 60 billion land and marine animals killed to feed Americans each year suffer throughout their lives, and some suffer further at the time of slaughter—in each case in measurable ways—perhaps assessments based on conventional economics omit a material component.

Furthermore, the estimate of $20.7 billion yearly covers only five inhumane practices. It doesn't include numerous others, like raising veal calves in crates, force-feeding ducks to produce foie gras, castrating pigs and cattle without anesthetic, and killing male chicks by starvation, suffocation, or grinding. Adding these and other practices to the calculation might double or triple the total.

What Now?

The living conditions discussed in this chapter and Appendix D might seem apocryphal, exceptional, or illegal. They're not. They're the normal, lawful, day-to-day conditions that industrially raised animals routinely face. For some, these images may suggest a need for change.

Want to have an immediate impact? Here's one idea: stop eating eggs and products made with eggs. Laying hens have it tough regardless of whether they're squeezed into battery cages, stuffed into enriched cages, or crammed into so-called cage-free buildings at unregulated densities. For a bird subjected to a painful debeaking, starved on a regular basis, and bled to death—while alert—eight years before her time, the differences between one kind of cramped living quarters and another are largely inconsequential. Besides, there's little to suggest that eggs are good for you and much to suggest they're not (*see* Appendix A). Giving them up will likely lower your cholesterol and could help prevent or reverse heart disease.

If you like eggs for breakfast, try a grilled tofu patty as a fried egg alternative—or a tofu scramble instead of scrambled eggs. For baking, try replacing eggs with banana, applesauce, or a commercial starch-based egg replacer. With these and other plant-based egg substitutes widely available, today it's easier to give up eggs than ever before.

Food for Thought

- Science has rejected the once-popular Cartesian view that animals don't feel pain or suffer. Today, we know that all animals raised or caught for food, including fish, feel pain, fear, and stress.

- In industrial animal production facilities, where virtually all US animal foods originate, animals are routinely subjected to pain and stress throughout their lives—and often at the moment of their deaths. The biggest problem in fish and factory farms is lifelong hyper-confinement and, for most animals, an almost complete inability to engage in natural behaviors. Routine, unanesthetized amputations of animals' body parts, including debeaking, castration, and tail docking, provide additional sources of pain and stress.

- Americans are overwhelmingly concerned about these practices and willing to pay to end them. Live auction research using real dollars suggests Americans would be willing to pay an average of $1,700 per person to end the most egregious of these practices. But until these practices are stopped, animal food producers will continue to impose an estimated $20.7 billion annually in externalized cruelty costs on US consumers.

9

Fishing Follies

One brilliant Alaskan day a few summers ago, I was sizing up the Kenai River and thinking about putting my kayak in the chilly water. I had driven up from California, paddling some of North America's great rivers along the way—the Skagit in Washington, the Bow in Alberta, and the same Yukon stretch that Jack London worked as a guide during the gold rush. But I hadn't seen anything like this in my travels. It was salmon season, and the fish were running. At the river's wide mouth, where the silty, gray waters emptied into the Pacific Ocean's Cook Inlet, the banks and shallows of the lower Kenai were lined with anglers by the score.

Salmon can be taken in that part of the river only by hand or net. And since most people lack the hand size or speed of a bear's paw, they use a hoop net—a nine-foot pole attached to a huge, netted bag. While I watched, people of all skill levels and walks of life yanked glinting king salmon from the water with little effort. Further upstream, where the water narrows, the spruces and willows grow denser, and people aren't allowed to fish, I would later see eagles and grizzly bears enjoying their share of the bounty. The regular glimpses of salmon being taken by people and animals suggested abundance almost without limits.

During its ocean phase, the royalty of the salmon family sports a sparkling silver coat with areas of blue-green or purple on its back. Once known as June Hogs because of their size and the seasonality of their runs, king salmon routinely used to weigh 80 to 100 pounds as adults and grow to five feet or more in length. But today, kings are lucky to reach half that size. As I later learned, the scenes that I mistook for the animal's abundance in Alaska were in fact images of decline. In fact, the king salmon fishery, like so many others on the planet, is in distress. The annual Alaskan catch of king salmon has fallen by half in the past several decades, and the numbers are even worse in Washington, Oregon, and California.

Beyond the obvious reason for this decline—overfishing—there are additional causes whose connections are harder to fathom. Salmon hatcheries, for example, are intended to help wild salmon populations. So why do hatcheries wind up harming the very species they're trying to help? The answers lie in the supply-driven pressures of meatonomics—which maybe, just for this chapter, we'll call "fishonomics." In the pages below, I explore these and other questions and suggest solutions to some of fishing's pressing economic problems. It should come as little surprise that these problems are largely about the money—that is, the desire of those who fish industrially or operate commercial fish farms to maximize their profits.

Incidental Taking

It's instructive to look at another marine animal whose numbers, like those of salmon, are shrinking: the leatherback turtle. The planet's fastest reptile, and the largest one after crocodiles, the leatherback has a ridged back and looks like an art deco spaceship in miniature. This prehistoric creature can grow to nearly ten feet, swim 20 miles per hour, migrate six thousand miles, and dive almost a mile. It's also unique among turtles for its lack of a bony shell, instead sporting a carapace of firm, rubbery skin.

Leatherbacks debuted in the Cretaceous period and have been around for 110 million years. But if things continue as they have for the past few decades, this century could be their last. Leatherback populations have fallen by as much as 95 percent in the past twenty years, and the species is now listed as critically endangered under the Endangered Species Act and international treaties.[1] These laws prohibit killing leatherbacks or selling their body parts. Yet despite this protection, tens of thousands of leatherbacks and other threatened or endangered sea turtles are *legally* killed yearly.

The loophole is a provision that permits what's known as incidental taking. This exception allows fishing enterprises that follow certain guidelines to lawfully kill endangered animals in the normal course of fishing. In federal waters, for example, shrimp trawlers can kill endangered turtles with impunity while fishing—provided they've installed a turtle excluding device (TED) in their nets, which theoretically allows turtles to escape. However, TEDs aren't perfect, and they often fail to save turtles' lives.[2] What's more, the use of TEDs is not monitored or enforced well in foreign waters, and they are not even required in many US state waters. Thus, in the Gulf of Mexico, where TEDs are not required, fishing activities in 2010 and 2011 led to an eightfold increase in sea turtle deaths over prior years.[3] "One of the greatest threats to sea turtle populations," notes the UN, "is capture in fishing gear."[4]

This massive collateral damage is a consequence of the fish industry's counterpart to land-based CAFOs: factory fishing. Today, more than twenty-three thousand factory ships weighing 100 tons or more patrol the world's oceans, typically staying at sea for weeks at a time and catching and processing huge quantities of fish.[5] Two of the most common industrial fishing methods, trawling and longlining, are also among the least discriminate. Trawlers drag a large, open-mouthed net that catches everything in its path. Longliners, by comparison, pull a length of individually baited hooks that trail for fifty miles or more behind the ship, enticing any hungry animal in the vicinity to take a fatal bite. Like trawl nets, longlines snare considerable amounts of unintended haul, or bycatch.

While some countries prohibit discards of certain species, the overwhelming practice is to throw back bycatch dead.[6] Thus, in most of the world's oceans, where little attention is paid to bycatch, factory fishing spells trouble for every fish, bird, mammal, or reptile unlucky enough to make contact with the juggernaut of nets or hooks that trail behind ships. Besides endangered turtles, other rare or threatened animals like dolphins and seabirds often die in a net or on a line. The albatross is one of the planet's most threatened creatures, with seventeen of twenty-two albatross species considered vulnerable, endangered, or critically endangered under international standards (the other five are labeled "near endangered").[7] One study counted at least forty-four thousand fishing-related albatross deaths in southern oceans each year, with researchers concluding that longlining was causing "serious declines in albatross populations."[8]

According to the latest estimate, the amount of marine life killed each year as bycatch is a stunning *40 percent* of the total worldwide intended catch.[9] That's almost 200 million pounds of nontarget, dead animals *each day,* or 10,000 pounds in the time it took you to read this sentence.[10] Whether you love seafood and eat it regularly or never touch the stuff, this arbitrary waste of life just reeks. As the authors of a 2009 study noted, bycatch is "one of the most significant nature conservation issues in the world today."[11] But even these massive figures don't tell the full story, as the bycatch ratios are much higher in certain regions or in connection with certain target species. Flatfish trawling in Alaska generates two pounds of bycatch for each pound of target fish.[12] Shrimp trawling in the Gulf of Mexico generates 10 pounds of bycatch for each pound of shrimp.[13] Further, the issues extend beyond the long shadow of dead animals trailing each plate of shrimp scampi or baked cod. Bycatch generates huge economic costs as well.

The hidden price of bycatch stems mainly from two problems: loss of juveniles of the target species and destruction of nontarget fish with commercial value. For example, researchers estimate that in the Northeast, eliminating bycatch would double the value of the Gulf of Maine fisheries, or regional fish habitats.[14] Another estimate finds that the value of marketable species discarded in the North Sea bottomfish fishery equals the value of fish caught.[15] In light of these and similar studies, the UN notes that aggregate annual economic losses resulting from bycatch "run into billions of dollars" and "in many fisheries the losses due to discard mortalities . . . *equal or exceed* landed catches."[16]

The loss of juveniles as bycatch is one of the reasons fisheries deteriorate and become underproductive. Nevertheless, destructive economic forces encourage commercial fishing fleets to keep operating even in overexploited, unprofitable fisheries. The World Bank and UN report that even as the world's fisheries continue to decline, twice the vessels needed for the task continue to pursue the global fishing catch.[17] With so many of the world's fisheries losing productivity, why do fishing enterprises keep exploiting them and making them even less profitable? The answer lies in a powerful economic force: fishing subsidies.

Subsidies on the High Seas

Give someone a fish, and you feed him for a day. Subsidize his fishing, and he'll do it for life—whether or not it's sustainable. In the United States, state and federal governments dole out $2.3 billion in handouts to the US fishing industry yearly.[18] That is to say, of the $0.59 average per pound collected by US fishers for their catch, $0.28 (or nearly half) is paid by taxpayers.[19] Yet the United States is not the only country to subsidize its fishing industry, and we're not nearly the most generous. Worldwide fishing subsidies total more than $25 billion, with Japan and China paying the most at $4.6 billion and $4.1 billion, respectively.[20]

Because fishing subsidies artificially reduce production costs, they encourage people to fish even in overexploited and unproductive fisheries. The World Bank and the UN say these incentives, which they condemn as "perverse," cost the fishing industry $51 billion yearly by diminishing the productivity of the world's oceans.[21] Subsidies also contribute to the collapse (that is, the decline by more than 90 percent) of many of the world's fisheries. Worldwide, humans have caused nearly one-third of fished species to collapse.[22] Without aggressive intervention, experts say, this trend will lead to the global ruin of *all fished species* by the middle of this century.[23]

When a regional fish population collapses, it not only idles humans employed in the local fishing sector, but it also disrupts the local marine ecosystem. "The least movement is of importance to all nature," remarked the polymath Blaise Pascal centuries ago. "The entire ocean is affected by a pebble." Research shows that the loss of biodiversity caused by one marine species's collapse makes it more likely that other species in the ecosystem will go bust.[24] It's a classic domino effect. A collapsed fishery typically leads to reduced populations of larger fish, marine mammals, and seabirds, and the loss of competition can cause an increase in populations of less marketable species like jellyfish.

A fishery's disintegration can also lead to bizarre and unexpected human responses. Commercial fishing extracted mountains of valuable cod from the North Atlantic for more than a century, taking well over a billion pounds in some years.[25] When overfishing led to the cod fishery's collapse in 1992, Canada reacted by banning cod fishing in the North Atlantic. Practically overnight, thirty-five thousand Newfoundland fishing workers lost their jobs.[26] The ban was unpopular, to say the least, especially among Newfoundlanders, and according to some commentators, the government needed a scapegoat.

Canada had formerly banned the hunting of another threatened species, the harp seal. While up to 9 percent of a typical harp seal's diet can consist of cod, the seal also helps the cod population by eating larger fish that prey on cod, like halibut, and accordingly is thought to have little overall effect on cod populations.[27] Nevertheless, several members of the Canadian government publicly blamed the harp seal for the cod fishery's collapse. In 1995, under the direction of Newfoundlander Brian Tobin, Canada's minister of fisheries and oceans, the country lifted the seal hunting ban. Canada now hosts the largest marine mammal hunt in the world, with an annual quota of 330,000 harp seals.

Fish Farming

Three billion people around the globe regularly eat what the French call, in something of a naïve misnomer, "fruits of the sea." As output from the planet's wild fisheries drops like a barometer before a storm, aquaculture—or fish farming—is increasingly taking up the slack. That makes it the planet's fastest growing food production system, and today, half the fish Americans eat is raised in tanks, cages, and other confinement systems.[28] Many believe aquaculture is a sustainable, cost-effective way to feed the planet, especially as the production of meat and dairy is increasingly seen as unsustainable at the levels the world demands. Is fish farming the way of the future?

Recall the Polyface Farm model for land animals. The technique of ecological rotation for farming livestock is one of the most sustainable ways (even if not *completely* sustainable) to feed animals, manage waste, and avoid degrading the land. In a similar vein, innovative fish farming methods surround the target species with a mini-ecosystem to promote natural waste management. Salmon, for example, might be raised next to shellfish, which filter solid waste, and seaweed, which processes nitrogen. One such efficient system is called Integrated Multi-Trophic Aquaculture (IMTA), and it's in use today at several aquatic farms in Canada's Bay of Fundy. IMTA isn't perfect, but it does help address one of the biggest concerns in fish farming—the effects of fish waste on surrounding ecosystems.

But waste is only one of several issues in aquaculture sustainability. In fact, the most eco-friendly way to raise fish is to grow them in land-based tanks or ponds where waste is completely contained, disease is minimal, and escapes are impossible. However, unlike open-water systems, land-based systems require costly processes to remove waste and to maintain water's salinity, temperature, and oxygen content.

Fish-farming scientists at the University of Maryland have sought to address the ecological limitations of fish farming. Led by Yonathan Zohar, chair of the university's Department of Marine Biotechnology, the group has developed a fully self-contained, land-based aquaculture system. The Maryland system uses bacteria to filter nitrogen from the water and microbes to convert waste to methane for use as a biofuel. It's about as eco-friendly as fish farming can be. However, it's not ready for production—and may never be. Not surprisingly, this system is limited by its high operating costs and the huge amounts of energy needed to run it.

Aquaculture methods like IMTA and the Maryland system are promising. But just as Polyface Farm is well-intentioned but ultimately unsustainable, these and other innovative fish farming methods also fall short. For starters, land-based systems, even those that don't take the costly extra step of recycling water and waste, are expensive. In a cage or pen system, by contrast, because the permeable container sits in open water, operators spend nothing to dispose of waste or to provide a constant supply of clean, oxygenated water. This fundamental difference allows cage and pen systems to operate much more cheaply than land-based systems. It also explains why cage aquaculture is the predominant method of fish farming throughout the world.[29]

IMTA, for its part, may be an effective way to protect ecosystems from fish waste. However, IMTA doesn't address a number of other ecological concerns associated with open-water fish farming discussed below. And even if they did, IMTA and similar systems are in use at only a handful of fish farms. The reality is that fish farm operators, like any business owners, look to the bottom line—and we know that price tags are closely watched in the world of meatonomics. That's why inexpensive cage-based systems are the standard. And as experience shows with largely futile efforts to reform land-based animal production methods, in an industry characterized by regulatory capture and heavy influence over lawmaking, it's likely to stay that way. Hence, in evaluating fish farming, while the experiments of innovators on the fringe are interesting, the relevant point of inquiry is the system as it exists today and is likely to persist in the future. And here's where the water gets a bit murky. Considering the many documented environmental impacts of aquaculture as practiced today in North America and around the world, the claim that it's sustainable emerges as something of a fish story.

Sustainability Issues in Fish Farming

Fish farms are the factory farms of the sea. And just like CAFOs, aquaculture relies on hyper-confinement to raise the largest number of animals in the smallest possible area. With two-thirds of the planet covered in water, it might not be necessary to stock fish as densely as battery hens. But necessary or not, that's how it's done. Foot-long trout, for example, are raised in densities as tight as twenty-seven to a bathtub-sized space.[30] And just as such tight densities cause problems in factory farms, they cause a variety of issues in fish farms.

Fish are susceptible to parasites. While these vermin can only achieve infestation at high density levels, a typical fish farm provides the Goldilocks-like stocking levels they need.[31] Atlantic salmon, the most common cage-reared fish, are particularly prone to sea lice. A parasitic crustacean measuring an inch or longer and resembling a miniature horseshoe crab, these dogged little creatures eat the blood, mucus, and tissue of living salmon. Because sea lice can only survive in saltwater, they typically drop off in the wild as their hosts migrate into fresh water. In saltwater fish farms, however, lice remain attached until removed by chemicals or, in some cases, gobbled by lice-eating cleaner fish.

A female sea louse lays up to twenty-two thousand eggs during her seven-month lifespan.[32] On tightly packed fish farms, newly hatched juvenile lice have little trouble finding a host to chew on. Picture, if you will, the huge numbers of concentrated salmon and egg-laying sea lice in a typical fish farm environment. With more than five hundred thousand salmon in an average farm, if just one in ten fish hosts just one female louse, and each louse lays just half her capacity, that's a localized plague of more than 500 million baby sea lice. Besides hurting farmed fish, these infestations also harm the surrounding ecosystem and its inhabitants. Like a swarm of tiny locusts, the hungry parasites explode into their surroundings and snack on any wild salmon in the vicinity.

Not surprisingly, sea lice from salmon farms are killing wild salmon populations.[33] On Canada's Pacific coast, for example, sea lice infestations are responsible for mass kill-offs of pink salmon that have destroyed 80 percent of the fish in some local populations.[34] But the damage doesn't end there, because eagles, bears, orcas, and other predators depend on salmon for their existence. Drops in wild salmon numbers cause these species to decline as well.[35]

Some farmers respond to lice by dosing the water with concentrated chemicals that kill the tiny creatures. Not surprisingly, adding toxins to the ocean harms the local ecosystem.[36] One study, for example, found that cypermethrin (used to kill lice on salmon) kills a variety of nontarget marine invertebrates, travels up to half a mile, and persists in the water for hours.[37]

But even more threatening to local ecosystems than sea lice and the chemicals that kill them are the massive quantities of waste generated by most fish farms. Consider aquaculture's effect in Scotland. In 2000, Scotland's fish farm industry created as much waste-based nitrogen as did two-thirds of the country's human population of 5 million—plus almost double the phosphorus that the human population generated.[38] Fish waste typically falls as sediment to the seabed in sufficient quantities to overwhelm and kill underlying marine life in the immediate vicinity and for some distance beyond. It also promotes algal growth, or the ironically named process of eutrophication (literally, "providing nourishment"), which reduces water's oxygen content and makes it less capable of supporting life. In 2008, the Israeli Ministry of Environmental Protection shut down two fish farms in the Red Sea that produced 2,000 tons of fish annually, because research showed that eutrophication from the fish farms was damaging the region's coral reefs.

Aquaculture also results in regular escapes by farmed fish into the world's oceans. In the North Atlantic region alone, up to 2 million runaway salmon escape into the wild each year.[39] The result is that at least 20 percent of supposedly wild salmon caught in the North Atlantic are of farmed origin.[40] Escaped fish breed with wild fish and compromise the gene pool, harming the wild population. Embryonic hybrid salmon, for example, are far less viable than their wild counterparts, and adult hybrid salmon routinely die earlier than their pure-bred relatives.[41] Like hitting a fighter when he's already down, this gene-pool degradation causes further declines in wild fish stocks that have already been pounded by overfishing.

Where Are the Little Fish?

Another direct and unfortunate consequence of aquaculture is over-fishing—the practice of capturing more fish than a fishery can regularly replace. Throughout the world, as fish farming explodes to meet surging demand, industrial fishing operations catch prey fish like anchovies, herring, and mackerel in increasing numbers to feed to captive fish. The top ten farmed fish consume an average of 2 pounds of wild fish for each pound they weigh at slaughter. The ratio is even higher for strictly carnivorous fish like salmon and tuna, which eat up to 5 pounds of wild fish per pound raised.[42]

The aquaculture industry's voracious appetite means that prey may soon join predators in the lists of overexploited and threatened fish stocks. Seven of the world's top ten fisheries are prey fish, with today's total catch in this category exceeding that of 1950 by a factor of four.[43] Most of the millions of tons of prey fish caught each year are consumed by aquaculture—the latest data show that farmed fish eat between 50 and 80 percent of all prey fish captured.[44] (Nearly all of the rest, by the way, is fed to pigs and chickens in factory farms.)

But wild animals have to eat too, and thousands of species depend on these little marine creatures. As stocks of prey fish decline, predator populations deteriorate in lockstep. In 2009, the nonprofit group Oceana released a report titled "Hungry Oceans," which highlighted the problem of dwindling predator populations. The report blames the depletion of prey fish stocks for declines in whales, dolphins, seals, sea lions, tuna, bass, salmon, albatross, penguins, and other species.[45] "We have caught all the big fish and now we are going after their food," said Margot Stiles, lead author of the Oceana report. The result, said Stiles, is "widespread malnutrition" in the oceans.[46] The main force behind this crisis, according to the report, is aquaculture's need for feed.

One way to reduce the quantity of fish used as feed in aquaculture is to raise herbivorous fish like tilapia. Unlike salmon, tilapia can thrive on pellets of corn or other grains. However, tilapia raised on this unnatural diet provide low levels of the omega-3 fatty acids that many people consider one of the main benefits of eating fish. That's because omega-3s originate in aquatic plants and are found in fish that eat those plants, but are not present in land-cultivated feeds. As a result, 3 ounces of farmed salmon typically contain more than 2,000 milligrams of omega-3s, while the same portion of farmed tilapia contains only 135 milligrams.[47] Moreover, research shows that farmed tilapia contains fatty acids at levels which increase the risk of heart disease. A 2008 paper published in the *Journal of the American Dietetic Association* found that omega-6 fatty acid levels in farmed tilapia were "so high they can be considered detrimental."[48] Another problem with tilapia is the African fish's propensity to invade lakes and displace native species through aggressive feeding and breeding. In fact, escapes from a tilapia farming operation in Lake Nicaragua, the largest lake in Central America, are blamed for the disappearance of the lake's less aggressive species like rainbow bass.[49] The upshot is even if tilapia might take some pressure off diminishing prey fish populations, the downside is considerable.

Yet perhaps the most surprising way that fish farms affect wild fisheries is through an economic phenomenon known as the Jevons paradox. William Jevons was a 19th-century economist who noted that as better technology made coal-burning equipment like steam engines more efficient, demand for coal counterintuitively rose instead of falling. That's because efficiency improvements lead to economic expansion and higher commodity sales, negating the effects of any efficiency per unit produced. Similarly, critics argue, as aquaculture makes fish production increasingly efficient, and fish becomes more widely available and less expensive, demand for fish increases across the board. This of course drives more fishing, which hurts wild populations.

Can higher production efficiency really spur greater demand for fish? Consider the change in worldwide salmon production and demand from 1987 to 1999. During this period, salmon hatcheries widely replaced natural salmon spawning areas lost to habitat destruction. As the result of lower prices and wider availability, world demand for salmon increased more than fourfold during this period.[50] Although hatcheries were expected to ease the pressure on wild salmon populations, the increased demand they caused actually aggravated the pressure.[51] Such supply-driven effects are common around the world, as increased availability and lower production costs—fueled by subsidies and industrial methods of fishing and fish farming—help lower prices and spur consumption.[52]

The Future of Fish Farming

The pressure on wild populations, the prevalence of parasites and disease, the lopsided input-output ratios, and the high levels of concentrated waste and chemicals all suggest aquaculture is hard-pressed to deliver on its promise to feed the world sustainably. In fact, these problems have led a number of scientists to condemn fish farming as irresponsible.[53] One group of researchers, for example, analyzed the practice of farming salmon and trout in Sweden's coastal waters. Taking into account its large externalized costs, the researchers concluded the system "is not only ecologically but also economically unsustainable."[54] Another team of scientists concluded that pen farming in China is an "economically irrational choice from the perspective of the whole society, with an unequal tradeoff between environmental costs and economic benefits."[55] Throw in the ethical issues associated with fish farming discussed in chapter 8, and it's evident that cage aquaculture—the world's predominant fish farming system—presents as many problems as land-based factory farming.

Conceivably, fish farming *could* be ecologically fixed by containing and greening it like at the University of Maryland. But as we've already seen, its high costs make this model commercially unviable. Just like CAFO operators, industrial fish farmers seek a regulatory framework that lets them externalize as much of their costs as possible. And they seek to operate in a way that complies with that framework at a minimum of expense. Can you blame them? Like most business owners, they're in it to make money—not to save the planet.

Around the world, regulation of fish farming is spotty and inconsistent. Since 95 percent of the farmed fish that Americans eat is imported, mostly from lesser-regulated regions like China and South America, most Americans who eat fish unwittingly skirt our own regulatory system. Jeffrey McCrary is an American fish biologist who has spent ten years studying how a small tilapia farm degraded Lake Apoyo in Nicaragua. "One small cage screwed up the entire lake—the entire lake!" McCrary said in a 2011 interview with the *New York Times*. "We wouldn't allow tilapia to be farmed in the United States the way they are farmed here," McCrary said, "so why are we willing to eat them? We are exporting the environmental damage caused by our appetites."[56]

The United States has, so far, trailed the rest of the world in aquaculture production. Half the fish we eat is farmed, but only 5 percent is raised here. But times are changing, and the federal government is starting to push fish farming almost as vigorously as it markets milk and meat. The National Oceanic and Atmospheric Administration (NOAA) urges, in a special area of its website dedicated to aquaculture, "A compelling case can be made for growing more seafood in the United States." That's because, among other things:

> The $1 billion value of total U.S. aquaculture production (freshwater and marine) pales in comparison to the $100 billion value of world aquaculture production. According to the latest information from the United Nations Food and Agriculture Organization, the United States ranks 13th in total aquaculture production.[57]

With this kind of government agenda, the United States certainly seems well-poised to improve its standing in the global farmed fish market. Of course, this prognosis should give you flashbacks to earlier chapters on land-based animal farming because, with little doubt, the producers who lead the charge will seek to dump most of their production costs on the rest of us.

Adding It Up

As with other methods of raising animals for food, fish production heaps huge external costs on society. Fishing's advocates say it provides jobs and benefits the world economy, but the UN begs to differ. In light of subsidies and other problems discussed already, the UN says, "Global marine fisheries . . . represent a *net economic loss* to society."[58]

Consider the externalized costs of aquaculture, which experts find can range from 20 percent to more than 100 percent of the value of a fish farm's output.[59] There are no published estimates of fish farming's external costs in the United States, but because research finds these costs often equal the value of fish produced, aquaculture's external costs can be conservatively pegged at half the total value of production, or $650 million yearly.[60] However, those are just the US costs. Since we import 95 percent of our farmed fish, almost all of the external costs of production are imposed on other countries.

Then there are the costs of overfishing. Recent research published by a team of experts suggests that overfishing in North America leads to losses of about 30 percent of the actual catch.[61] Applying this estimate to the portion of total US landings ($4.8 billion) indicates that the annual US cost of overfishing is about $1.4 billion.[62] But again, this figure neglects the imported fish Americans consume. Also, this limited estimate fails to account for hundreds of nontarget species deprived of food, most of which have no recognized economic value, whose numbers are dropping around the world as overfishing starves them out of existence.

Finally, there are the costs of bycatch. As we've seen, research shows that bycatch losses often *equal* the value of the total landed catch. While there are no definitive figures for the United States, the costs of US bycatch losses can be roughly estimated at half our commercial landings of $4.8 billion—that is, $2.4 billion.[63]

CHART 8.1 Annual Externalized Costs of US Fish Production (in billions of dollars)

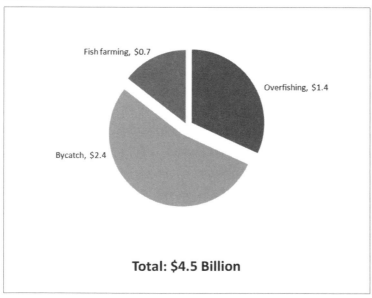

Fish farming, $0.7

Overfishing, $1.4

Bycatch, $2.4

Total: $4.5 Billion

What's Missing?

The $4.5 billion in estimated externalized costs of US fishing activities amounts to more than 70 percent of the US fishing system's total output of $6.1 billion.[64] Oddly, compared to the hundreds of billions in costs that land-based animal agriculture generates, this doesn't sound like much. And in fact, because this figure is limited to US costs, it's deceptively small. But then, most of the external costs of producing fish for American consumption are imposed on people outside the United States.

There are other significant, but unavoidable, omissions from this estimate. Oceans supply a variety of indispensable ecosystem services to the planet—life-support systems that keep humans and other animals alive and fed. Among others, these include filtering services that keep water clean and nursery habitat services that provide important areas where juveniles can develop (such as wetlands, oyster reefs, and sea grass beds). A 1997 study estimated the total value of marine ecosystems services was $21 *trillion* in 1994 dollars—or $33 trillion today.[65] To put this in perspective, US GDP—the highest in the world—is about $15 trillion.

Because of the interconnectedness of species in an ecosystem, when ocean biodiversity declines, a marine ecosystem's ability to provide these valuable services also falls. Since 1800, as human activity has reduced marine biodiversity, filtering and nursery habitat functions have each fallen by more than 60 percent.[66] The economic costs of drops caused by fishing in these and other ecosystem services have not yet been measured. But when these costs are calculated, and presumably someday they will be, they could easily measure an order of magnitude more than the $4.5 billion we can calculate today.

Rethinking Supply and Demand

Is the supply-driven focus on ramping up fish production misplaced? Fishers and fish farmers are decimating the world's oceans to catch or raise fish in ever greater quantities, but the planet's fisheries should now be exploited *less*, not more, if they are to recover and heal. And the evidence shows that fish farming—that reputed paragon of food production—just makes things worse. Among other failures, we've seen that aquaculture causes many of the same environmental problems as factory farming. Further, aquaculture's practice of robbing Peter to pay Paul—overexploiting little fish to feed bigger ones—just ratchets up the pressure on depleted fisheries and threatened species.

Given the earth's limited resources, maybe the answer to the fishing question isn't to crank *up* supply—but rather to turn *down* demand. As with meat and dairy, demand for fish has been artificially inflated by economic forces, and the world simply doesn't need to eat fish at these—or any—levels. One important path toward restoring true market equilibrium is to eliminate fishing subsidies in the United States and around the world. And on a personal level, you might consider replacing some or all of the seafood in your own diet with plant-based proteins. Omega-3s are readily found in a variety of vegetable sources—including flax, hemp, soy, and walnuts. And unlike virtually all fish, these sources *don't* contain cholesterol, mercury, and PCBs (*see* Appendix A). Consider the consequences of just giving up shrimp. With the highest bycatch-to-target ratios in the industry, a few plates of foregone prawns could save a dozen other fish from the discard pile.

Some believe the answer is to stop eating seafood altogether. After all, considering the harmful environmental effects of either catching or farming fish, it's sort of a damned-if-you-do, damned-if-you-don't dilemma. Maybe that's what George W. Bush had in mind in 2000 when he astutely observed, "I know the human being and fish can coexist peacefully."

Food for Thought

- Industrial fishing and aquaculture activities are damaging the world's oceans. If not curbed, factory fishing operations will eliminate all currently fished species by the middle of this century. While fishing's proponents defend the industry on the grounds that it creates jobs and helps the world economy, the

UN complains that on the contrary, fishing represents a net economic loss to the world.

- Although we import most of our seafood, the economic costs to Americans of US fish production are nevertheless sizable. Fishing's biggest economic problem is that of bycatch, or incidental taking, with nearly 200 million pounds of marine life killed and discarded each day in the world's oceans. In the United States alone, bycatch costs an estimated $2.4 billion yearly. Overfishing is another leading source of destruction and economic loss, costing US taxpayers and consumers roughly $1.4 billion each year. And fish farming, with its heavy environmental damage and destructive pressure on wild populations, is both unsustainable and socially expensive—with a cost to Americans of nearly $1 billion in externalized costs each year. Yet because most of our fish come from outside the United States, the $4.5 billion in total US externalized expenses represents just a tiny fraction of the worldwide hidden costs related to Americans' fish consumption.

- As the Jevons paradox predicts, while fishers and fish farmers catch or raise marine life in increasing quantities, the higher availability and lower production costs of seafood (stemming from subsidies and industrial production methods) keep retail prices low. These supply-driven forces push fish consumption higher, and they accelerate the destruction of the world's oceans. It seems the only way to find a sustainable equilibrium between supply and demand for seafood is to lower demand. On an individual level, that might mean making a personal decision to eat less fish. And at an institutional level, it means the federal government must end fishing subsidies and adopt policies to lower demand—as discussed in the next chapter.

Recipes for Change

Years ago, I had a legal client I'll call Frank. Like a squirrel with a nut, Frank had a tenacious and single-minded focus on his company's bottom line. Once, early in our relationship, he handed me a check with such evident annoyance that I actually tried to make my later requests smaller. Another time, I suggested paying modest severance to a fired employee, although the company had no legal obligation to do so. As he did from time to time, Frank reminded me—in a joking tone that didn't hide the seriousness of his message—"Remember, we're in it for the money." In other words, despite an occasional show of concern for employees, customers, and other stakeholders, the company really existed for one purpose—to make money for Frank and a few other shareholders. If this management style seems mercenary, it isn't—nor is it atypical. It is, more or less, business as usual for just about any US company. "American corporations," observed historian Stephen Ambrose, "hate to give away money."

We've seen over and over in the preceding chapters that big business is in it for the money. But the consequences of profit-driven business practices go beyond the purely financial. This book seeks to quantify the hidden pecuniary effects of meatonomics, but numbers tell only part of the story—the real results go much deeper. When we lose a loved one to cancer or stroke, we suffer more than the costs of treatment or lost earnings. When we're bothered by the mistreatment of animals, it's not because we wonder how much we'd pay to end it but because it makes us question our own humanity.

In a sense, the dollar values in this book are mere *proxies* for social ills. They're measuring tools. The main problems are the underlying issues themselves, and those are the ailments this book seeks to address. This week, thousands of Americans will die from diseases related to eating animal foods. At any moment, 1 billion US farm animals are hyper-confined in stressful conditions, and groundwater sources in one-third of US states are polluted with animal waste. Although these problems' price tags are huge, dollars are just one way of measuring the damage.

In this final chapter, I propose steps Americans can take—both individually and collectively—to alleviate the social harms caused by meatonomics. That means improving our health, fostering ecological well-being, and lowering the system's costs—both economic and otherwise—to all of us. It's a complex set of issues, and although this book views them through an economic lens, the human and humane problems—not the cash—are the real concerns. Unlike corporations, living beings are in it for a lot more than the money.

Summary

Let's start with a recap of this book's major points. First, under ever-increasing pressure to raise campaign funds so they can get elected or reelected, lawmakers reward industry donors by passing legislation that surrounds meat and dairy producers with a precious cocoon of protection. These measures include ag-gag laws, cheeseburger laws, food defamation laws, customary farming exemptions, and animal enterprise terrorism laws. Although couched in terms of defending citizens, such protections routinely favor industry over consumers. These laws make it increasingly hard to investigate, criticize, or sue factory farmers over food safety, personal injury, or animal cruelty. This legislation is also part of a legal framework that permits producers to offload most of their costs onto society. "Big business is not dangerous because it is big," said former President Woodrow Wilson a century ago, but because of "privileges and exemptions which it ought not to enjoy."

Second, regulatory agencies frequently favor animal food producers over consumers. The USDA promotes meat and dairy highly effectively, using government messaging to bombard consumers in a variety of media. But when it comes to acting on its stated mission to protect us through food labeling, nutrition advice, and food safety programs, the USDA's loyalty to industry often results in confusing, misleading, and ineffective regulatory measures. The FDA has similar conflicts of interest, rendering the agency unable to withdraw animal drugs that even its parent, the Department of Health and Human Services, acknowledges are dangerous. It's become almost what former President Rutherford B. Hayes feared in 1876: "a government of corporations, by corporations, and for corporations." The animal food industry's capture of the agencies meant to regulate it is another support beam in the structure that lets industry impose its production costs on taxpayers and consumers.

Third, animal food producers engage in a sophisticated public relations campaign to convince consumers that meat and dairy are healthy in almost any quantity and that animals are treated well. However, a deeper look at these communications—and the underlying research—finds that much of what the industry tells consumers is inaccurate and misleading. In fact, the research overwhelmingly shows that meat and dairy are unhealthy, particularly in the supersized portions Americans consume them. Moreover, contrary to industry assertions that zero-grazing dairy cows are happy and that hyper-confined pigs love their conditions, the body of evidence shows something radically different. In fact, virtually all farm animals in the United States are raised in conditions of extraordinary stress and suffering, including routine, painful mutilation or amputation of body parts, and a lifelong denial of basic and natural behaviors.

This trifecta of lopsided lawmaking, regulatory failure, and industry doublespeak has saddled America with a set of massive, far-reaching problems in many categories. Americans are caught under the thumb of the supply-driven forces of meatonomics, and our ability to make informed, healthy choices about what we eat is heavily impaired. This combination of misinformation and artificially low prices fosters unnatural and undesirable levels of consumption. We eat more meat and dairy than any other people on the planet, and we pay for our high consumption with some of the world's highest rates of obesity, cancer, and diabetes. This gluttony hurts nature as much as it does us. Most US farmland produces below capacity because of soil erosion related to animal agriculture, and thirty-five thousand miles of American waterways are polluted with animal waste. Our surging demand for fish has destroyed fisheries and marine ecosystems in North America and the rest of the world.

Most of us sense that something is rotten in the state of meatonomics. Polls regularly show that voters are outraged by farm subsidies and consumers are alarmed by inhumane and unsanitary conditions on factory farms. Nevertheless, there is little change in the status quo, and most of us go about our daily lives—working to pay our share of the massive financial costs that these problems cause. As we've seen, the hidden expenses of meatonomics total more than $400 billion dollars yearly. Let's revisit those costs for a moment.

Contemplating the Costs

The hidden costs associated with animal foods are about the same as all state and federal spending on Medicaid, or about $20 billion more than Russia's annual government budget. However you look at this number, it's huge. For a line-by-line look at how this $414 billion estimate is calculated, *see* Appendix B.

CHART 10.1 External Costs of US Animal Food Production (in billions of dollars)

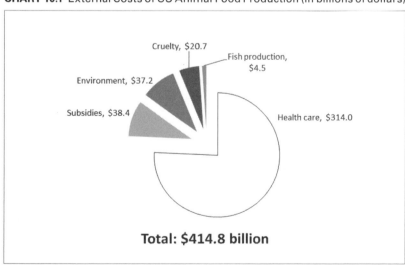

Cruelty, $20.7

Fish production, $4.5

Environment, $37.2

Subsidies, $38.4

Health care, $314.0

Total: $414.8 billion

Some may question the underlying assumptions and conclude that this huge total is overstated. But rather than being on the high side, this total likely understates the actual costs for several reasons. In order to keep this figure credible and focused on the United States, I've omitted data that can't be measured (but nevertheless exist), data that are irrelevant to the American economy (but relevant elsewhere), and data that standard economic theory does not currently recognize (but arguably should, and may someday).[1] Big problems call for big solutions, so the real question is, what can be done?

The Solution

I propose a three-part solution to the problems of meatonomics. First, adjust taxes to make animal foods more costly and to put cash in taxpayers' pockets. Beyond the financial boost, this change would give consumers more accurate price signals and lead to an important shift in consumption patterns. Second, restructure the USDA, clarifying its purpose to ensure that industry influence is minimized and consumers receive accurate information. And third, adjust federal support programs to reduce spending and better align financial support with policy goals. All told, these changes would lower the hidden costs of animal food production by an estimated $184 billion per year, save more than 172,000 American lives yearly, and cut our carbon-equivalent emissions by almost half each year. If it doesn't completely fix the failed market for animal foods, it's a big start with a serious impact. But before we get to the macro plan, let's start by laying out a few simple changes that anyone can make personally. To tweak an old adage: fixing the animal food industry starts at home.

Be the Change

"They always say time changes things," said Andy Warhol, "but you actually have to change them yourself." In fact, anyone can make an instant, personal change that will immediately reduce the burden of meatonomics on people, animals, and the environment. That simple shift is to consume less animal foods—or give them up altogether.

Want to lose weight? Studies find that on average, people on a plant-based diet weigh significantly less than omnivores. A 2009 research paper published in *Diabetes Care*, for example, found that after adjusting for confounding lifestyle factors like physical activity and television watching, vegans and meat-eaters had a mean body mass index (BMI) of 23.6 and 28.8, respectively (anyone with BMI over 25 is considered clinically overweight).[2] This finding led the study's authors to note the "substantial potential of vegetarianism to protect against obesity."[3] It's hard to argue with the 18 percent difference in BMI between the two groups. That means that if diet were the only factor in weight loss, an omnivore who weighed 180 pounds could shed 32 pounds by going vegan.

Want to lower your cholesterol? Studies routinely find herbivores have far healthier cholesterol levels than omnivores. One synthesis of published papers on the subject finds that American vegans and omnivores have average blood cholesterol of 146 mg/dl and 194 mg/dl, respectively.[4] These figures are particularly important because, while the magic number 200 mg/dl is often cited as the safe level for total blood cholesterol, considerable evidence points to the much lower 150 mg/dl as the truly safe level. The Framingham Heart Study, for example, is a long-running cardiovascular study on thousands of residents (spanning three generations) of Framingham, Massachusetts. Among study participants who suffered heart attacks, more than one-third had cholesterol levels between 150 and 200 mg/dl.[5] On the other hand, no one in the Framingham study with cholesterol below 150 mg/dl ever suffered a heart attack.

If you're not ready for wholesale change, consider some of the effects of skipping animal foods just one day a week. A single American's decision to forego meat, eggs, and cheese each Monday would cut 450 pounds from the total volume of animal waste generated for the year—enough to fill a bathtub to the brim. That decision would also spare at least five land animals each year from hyper-confined lives in factories. If a family of four skips steak one day a week, according to a 2011 report by the Environmental Working Group, the reduction in carbon emissions would be like taking a car off the road for three months.[6] There are also major health benefits in bypassing animal foods just one day a week. According to Allison Righter, director of the Johns Hopkins Meatless Monday Project, replacing four servings of red meat each Monday with nuts, legumes, or whole grains reduces one's risk of death by more than 11 percent.[7]

Not long ago, nuts and other stereotypical squirrel food would have been at the top of the list as substitutes for animal-based meat. But the market for vegetarian foods is growing fast, and in the past decade, food makers have introduced dozens of tasty, healthy, plant-based meats. Today, most grocery stores carry animal-free versions of foods like salami, bologna, pepperoni, pastrami, sausage, turkey, hamburger, tuna, shrimp, calamari, meatballs, corned beef, shredded beef, barbecued ribs, hot dogs, and chicken nuggets, to name a few. If you like the taste of meat but suspect you eat more of it than you should, it's easier than you might think to eat less or give it up completely.

The same goes for eggs and dairy. Plant-based substitutes abound for milk, cheese, yogurt, butter, ice cream, and eggs. Interested in cheese that's eco-friendly, humane, and cholesterol-free? Increasingly, markets carry plant-based versions of cheddar, jack, mozzarella, feta, blue cheese, cream cheese, and others. Instead of cow's milk, there's milk from oats, rice, hemp, soy, almonds, and coconuts (my own favorite is vanilla-flavored almond milk). Another humble recommendation: coconut-milk ice cream (it comes in flavors like coconut almond chip and chocolate peanut butter swirl). Plant-based foods don't taste exactly like their animal-based counterparts, and they may take a little getting used to. But then, maybe that's part of the point.

We vote with our pocketbooks every day, and our consumption choices matter. We've seen that animal food producers use a variety of techniques to make us buy their goods in the highest quantities possible, including keeping prices low to boost demand and disseminating misleading or confusing marketing messages. Perhaps a newfound understanding of the various insidious factors that influence our buying decisions will help readers make better-informed choices. Yet, while it certainly helps for individuals to spend their dollars outside of the animal food system, society must go even farther and reform the institutions that define this system.

Some will say it's naïve dreaming to suggest changes to one of the most powerful and intractable industrial complexes on the planet. Others have proposed reform, and others will, following the release of this volume. But maybe the constant reminders are helpful. Maybe it helps to have a working blueprint. Maybe it helps to catalog the benefits that changing the status quo can yield. At any rate, many believe society will soon be ready for changes like these—perhaps a lot sooner than we think. As civil-rights-activist-turned-congressman John Lewis asked, "If not us, then who? If not now, then when?"

Pigs, Pigou, and the Pigovian Tax

The same economist who introduced externalities explained how to fix them. Arthur Pigou, a Cambridge University professor active in the first half of the 20th century, conscientiously objected to World War I and drove an ambulance rather than serving in the military. He was also a famous intellectual adversary of economist John Maynard Keynes, although their public disagreements were offset by a strong private friendship. Between ambulance trips, Pigou found time to theorize that negative externalities can be fixed by taxing the goods that cause them. A Pigovian tax adjusts the market for goods that cause externalities by raising the goods' costs and thereby reducing demand for them. Such a tax pays a double dividend by both generating revenue and reducing undesirable consumption. Ideally, the new revenue adds to—or replaces—general tax revenue and thus can be used to lower general taxes.

Taxing authorities around the world have successfully used Pigovian taxes in a variety of ways. Most notable, perhaps, are programs that use tobacco taxes to reduce cigarette smoking and increase tax revenues. In the United States, we tax cigarettes at the federal, state, and—in some cases—city level. This combination is highest in New York City, where a pack of cigarettes carries taxes of $5.85 and sells for $11 or more. Amazingly, this hefty tax still falls far short of the total externalized costs of cigarettes. Recovering that entire figure would require imposing taxes of $10.47 per pack (according to the latest estimate by the US Centers for Disease Control and Prevention), or nearly three times the average pre-tax price of a pack of cigarettes.[8]

Nevertheless, cigarette taxes do work. Studies show that depending on the affected consumer group's age, income level, and extent of addiction, a 10 percent cigarette tax lowers smoking by 4 to 14 percent.[9] Moreover, cigarette taxes lead to huge revenue boosts. In Texas, for example, a 2007 cigarette tax hike of $1 per pack increased state tax revenues in the first year by more than $1 billion—while lowering cigarette sales in the state by 21 percent.[10]

Consider the huge gains that cigarette taxes have yielded for both personal health and government revenue since the early eighties, when state and federal governments began aggressively increasing these levies. From 1982 to 2007, inflation-adjusted prices of cigarettes nearly tripled as a result of increases in state and federal tobacco taxes.[11] During the same period, moving virtually in lockstep with the tax increases, US per capita cigarette consumption dropped more than 50 percent and the incidence of lung cancer among Americans fell 30 percent.[12] The tax increases paid a double dividend as well: despite the significant decline in consumption, combined state and federal tobacco tax revenues grew during the period from $15.7 billion to $25.2 billion (in inflation-adjusted dollars).[13] Table 10.1 shows the direct, negative correlation between cigarette prices and consumption, and table 10.2 shows the double dividend: an increase in tax revenues and simultaneous decrease in lung cancer incidence.

Taxes aren't the sole reason for Americans' declining consumption of cigarettes. For decades, government agencies and nonprofits have engaged in regulatory and public relations campaigns against tobacco use. These have yielded stricter labeling laws, restrictions on cigarette advertising, and billboard, radio, and TV ads that warn of smoking's dangers. But data from other countries, where cigarette taxes have been overwhelmingly successful even in the absence of American-style regulatory and public relations efforts, show that taxes are the most important part of any antismoking campaign. In Mexico, new tobacco taxes imposed between 1981 and 2007 led to cigarette prices tripling and consumption dropping by half.[14] In France, a 27 percent tax increase between 2003 and 2004 caused consumption to fall 7 percent.[15] And in Japan, a 2010 tax increase led to the steepest year-over-year consumption drop (2.2 percent) ever seen in that country.[16] These results from around the world show that even without help from public relations, taxes unquestionably work to reduce consumption.

TABLE 10.1 US Cigarette Prices and Consumption, 1982–2007

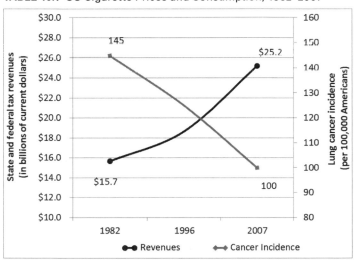

TABLE 10.2 US Cigarette Tax Revenues and Lung Cancer Incidence, 1982–2007

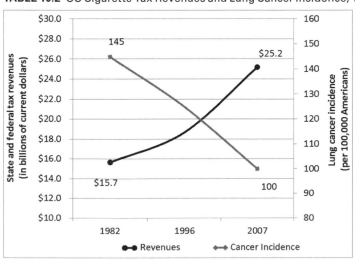

Taxing Meat and Dairy

Just as it has for cigarettes, a tax on animal foods would pay a double dividend by simultaneously boosting revenues and lowering consumption (and related social problems). Recall that the weighted average elasticity rate (that number that ties demand to price) for all animal foods is about 0.65.[17] A 3 percent tax on animal foods, for example, would reduce consumption by about 2 percent and generate about $7.4 billion in revenue.[18]

Of course, because we've seen that animal foods generate external costs worth nearly double their retail prices, a 3 percent tax in this category is unlikely to do much. The average cigarette tax in the United States is a hefty 72 percent.[19] Accordingly, in light of the slightly smaller ratio of external costs generated by animal foods, I propose a *50 percent federal excise tax* on all domestic retail sales of meat, fish, eggs, and dairy.[20] For simplicity, I refer to this across-the-board tax on animal foods as the Meat Tax.

The way the tax would work is simple. Every food item intended for human consumption that contains any animal product as an ingredient or component would be subject to a 50 percent tax imposed at the retail point of sale. For example, including the tax, a $10 store-bought steak would cost $15, a $4 Big Mac would cost $6, and a $2 Baskin-Robbins ice cream cone would cost $3. Goods containing only negligible amounts of animal foods, like cookies containing egg, or bread containing whey (a milk by-product), are nonetheless taxed at the usual Meat Tax rate. Why? Two reasons. First, trying to set thresholds for the tax's applicability would be both costly to administer and impractical to enforce. Second, any ingredient that is negligible can, by definition, easily be replaced or eliminated, which means producers who want to avoid the tax can easily change their foods' composition. The fact that, for example, many brands of bread and cookies contain no animal products shows how easy it is to make such items without animal foods. In fact, one of the tax's benefits is that it would encourage producers to reformulate the composition of such hybrid goods. As the proverb goes, "Necessity is the mother of invention." Plus, beyond healthier new edible combinations, the revised goods would impose lower external costs on society.

A Tax Credit in Every Pocket

"I shall never use profanity," said Mark Twain, "except in discussing house rent and taxes." Like Twain, many will find it hard to swallow a new tax, especially one applied to common dietary staples. That's why there's another feature meant to soften the blow.

In conjunction with the Meat Tax, each American taxpaying individual or family would get an annual tax credit averaging about $560 (the precise amounts range from $410 for single filers to $750 for married joint filers).[21] This would cost the US Treasury $78 billion, admittedly a nontrivial figure.[22] But the revenue and cost savings resulting from the Meat Tax will not only offset this cost but will also provide a comfortable cushion above it. Thus, after accounting for the reduced demand that it causes, the tax and other changes would yield an annual cash surplus of more than $32 billion.[23]

Why a tax credit? For one thing, the amount of the credit will offset the extra tax burden on each taxpaying individual or family (after adjusting for lower consumption), so that Americans' ability to eat will not be diminished. Further, the spending-related stimulus will help offset the lower spending caused by other parts of the plan. Finally, it's the quid pro quo, the benefit to individuals, that will motivate voters and lawmakers to support the overall plan.

Wouldn't a tax credit be useless to people who pay no income taxes—a group which recent estimates put at nearly half the US population? Actually, even those who pay no taxes can turn a tax credit into cash by simply filing a tax return. That's why some filers get tax refunds based on items like the Child Tax Credit and the Lifetime Learning Tax Credit even though they have no taxable income and pay no taxes. The fact that those with lower incomes are disproportionately more likely to pay no taxes *does not* mean the poor will be less able than others to take advantage of the proposed tax credit. To the contrary, there's evidence the poor are generally well-informed about tax credits and other benefits to which they may be entitled.[24]

Revamping the USDA

The USDA, as we've seen, is riddled with built-in conflicts that make it almost impossible to discern its purpose or message. Many critics, including lawmakers like former US Senator Peter Fitzgerald (R-IL), have proposed reorganizing the agency to eliminate some of these inherent contradictions. The USDA should keep its original, Lincoln-era purpose of helping farmers and other rural Americans. But the duties of regulating animal food labels and inspecting meat and dairy plants, which frequently require the agency to act in a way opposed to its clients' interests, must be revamped. The FDA already performs these functions for most other foods and is the logical choice to take over these tasks.

The USDA should also completely exit the business of promoting animal foods and leave that to the trade organizations. There is simply no reason for the US government to encourage the heaviest people on Earth to eat more fat-rich food. Checkoff programs for animal foods should be discontinued so costs are no longer routinely passed on to consumers for messages that many consider insidious and inappropriate. Some people argue that food is a necessity and this warrants government reminders to eat. Clothing is also a necessity, but Americans manage to dress ourselves without bureaucrats reminding us to buy and wear garments. We don't have—or need—a taxpayer-funded Department of Clothing to help Gap and Benetton sell their products. Because we've seen checkoffs drive about $4.6 billion in annual sales of animal foods, or 1.8 percent of the industry's total annual sales of $251 billion, we can expect that eliminating these programs would reduce consumption by about 1.8 percent.[25]

Finally, both the USDA's food stamp program, and the agency's authority to make nutrition recommendations, should be turned over to the Department of Health and Human Services. For decades, animal food producers have sought—usually with great success—to influence these programs in order to sell more goods. Removing the functions of nutrition assistance and guidance from the agency tasked with helping meat and dairy farmers sell more product would end the worst conflicts of interest that plague the USDA.

Hunting the Bounties

We're not done with the USDA. With annual spending on farm and nutrition programs of $209 billion, mostly funded by taxpayers, the agency is the primary vehicle through which federal farm subsidies flow.[26] It's also the best place to look for ways to reduce and refocus government largesse.

Not all welfare programs lack merit. The USDA's $115 billion Food and Nutrition Service, for example, helps farmers by boosting sales of crops and animal foods. It also helps one in seven Americans by providing school lunches and food stamps for low-income children and adults. There's no reason to eliminate these food assistance programs, but like alcohol and tobacco, animal foods should be excluded as ineligible for purchase. This change would shift demand in the low-income demographic—which is particularly at risk for obesity and diabetes—to healthier, lower-fat protein sources. It would also cut indirect support to animal food producers by $36.8 billion and reduce US sales of animal foods by $24.8 billion.[27]

The other proposed support change would better align the USDA's spending programs with its own nutritional recommendation that we eat less animal foods. Five USDA programs spend nearly $30 billion yearly supporting US farmers with loans, research, crop insurance, marketing assistance, and help exporting their products.[28] By eliminating the portion of this total spent supporting animal agriculture (i.e., subsidies to growers of livestock and feed crops), we can cut $18.2 billion from the subsidy pool.[29]

The reduced demand for meat and dairy caused by a Meat Tax would automatically drive further cuts in subsidies to the animal food industry. For example, state and federal irrigation subsidies pay farmers $15.7 billion yearly to water feed crops. As a result of the decline in consumption, roughly $5 billion of this total is likely to migrate away from supporting animal foods.[30] In total, as we'll see, the Meat Tax and other proposed changes would cut US consumption of animal foods by nearly half, save roughly $184 billion dollars in externalized costs, create a $32 billion cash surplus, shift tens of billions of dollars in support payments to healthier foods, and provide enormous other social benefits.

The Meat Tax Delivers

Consider the benefits. The Meat Tax and other proposed changes would reduce US consumption of animal foods by about 44 percent.[31] Further, the tax credit would help ensure that those who consume meat, eggs, fish, and dairy at the lower levels driven by the tax have enough extra cash both to pay for those goods and to pay the tax. Most important, the shift in consumption would help Americans in dozens of ways—some economic, and some that go beyond mere dollars. Assuming that cutting consumption results in pro rata declines both in production and in the problems that production and consumption cause, we can expect results like these:

- 172,000 fewer annual deaths from cancer, diabetes, and heart disease.[32]

- 26 billion fewer land and marine animals (eighty-five fewer animals per American) killed each year.[33]

- A 3.4 trillion-pound annual reduction in the emission of carbon dioxide equivalents.[34] (That's like garaging, each year the tax proposal is in effect, all 250 million registered highway vehicles *and* all twelve million motor boats and ships in the United States.)

- 440 billion pounds less hazardous waste generated yearly and available to enter American soil or water.[35]

- 708,000 square miles of US land no longer devoted to raising livestock or feed crops. (Subtracting the fraction of this land necessary to grow plants instead of animal foods leaves 566,000 square miles—an area twice the size of Texas.[36] In other words, one-sixth of the contiguous United States would no longer be needed for agriculture and could be returned to forest, grassland, or other native habitat.)

- $26 billion in annual savings to Medicare and Medicaid programs.[37]

- Annual decline of $184 billion in animal foods' external costs imposed on Americans.[38]

- Yearly cash surplus of $32 billion in the US Treasury.[39]

How would we spend such a huge annual cash surplus? Here's one idea: help fund a change in business model for farmers who raise animals or feed crops. Congress did just that for the tobacco industry in 2004 when, in conjunction with halting price supports, no-recourse loans, and other federal subsidies to tobacco growers, it set up the $10 billion Tobacco Transition Payment Program. Tobacco-growing states offered their own buyout programs as well, and thousands of tobacco farmers were able to transition to new crops. We might also use some of the surplus cash to create a permanent fund to clean up soil and water damaged by animal agriculture. Or to support organic farmers, provide Americans with better nutrition education, or pay down some of the national debt.

Trade-Offs

Is it too good to be true? As economist Thomas Sowell notes, "There are no solutions, only trade-offs." Policy changes invariably come with strings attached, and the Meat Tax is no exception. Consider some of the trade-offs and critiques

The tax is regressive—that is, it disproportionately affects the poor.

In fact, meatonomics already drives overconsumption and health problems that unfairly hurt the poor. For example, one cohort study finds that Americans in the lowest income group are consistently more overweight than those in the highest income group.[40] Another finds that low income is associated with a higher prevalence of diabetes.[41] Animal foods aren't the sole source of these health problems, but as we saw in chapter 6, they're a significant contributing cause. And by shifting demand among the poor away from animal foods and toward healthier protein sources, the Meat Tax will help ameliorate some of the health disparity between socioeconomic groups. Also, as noted, the proposal keeps food stamp program spending at current levels, and the $32 billion cash surplus it generates can provide additional assistance to low-income Americans.

It's too extreme.

Is it? The proposed demand cut seeks to return Americans' per capita meat consumption to its level in 1940, before the meteoric rise of fast food and factory farms—and before consumer demand for animal foods doubled. The incidence of diabetes was then one-eighth of today's level, and the obesity rate was one-third what it is now. Is it extreme to return to healthier consumption levels? In the documentary *Forks over Knives,* heart surgeon Caldwell Esselstyn responds to the argument that shifting consumption away from animal foods and toward plants is extreme. "Half a million people a year will have their chests opened up and a vein taken from their leg and sewn onto their coronary artery," notes Esselstyn. "Some people would call that 'extreme'."

It's not extreme enough.

Some may argue the Meat Tax does not go far enough to end the abuse of animals used for food—or that it merely preserves an inhumane status quo. For example, law professor Gary Francione has argued that those concerned for the well-being of animals must reject "new-welfarist" measures—halfway steps that purport to help animals but in fact provide little real benefit. According to Francione, such measures fail because they (1) support and validate the existing system, (2) cause little change in consumer behavior, (3) increase profits to producers, thereby encouraging further animal abuse, and (4) have little measurable, empirical effect on animal welfare.[42] However, these objections do not apply to the Meat Tax.

On the contrary, the proposed tax: (1) does *not* support the status quo—rather, it seeks to dismantle and repurpose nearly half of the animal food production system to plant-based foods; (2) would cause a massive change in consumer behavior, namely, a 44 percent drop in consumption of animal foods; (3) would significantly reduce animal food producers' viability, forcing many to exit the business; and (4) would have a major, measurable effect on animal welfare by saving the lives of 26 billion land and marine animals yearly. Unlike the new-welfarist measures that Francione and others reject, this proposal will achieve major changes to the existing system and tangible, significant benefits for animals.

The tax won't generate the revenue projected because producers will internalize it.

If producers want to reduce or eliminate a tax's effect on demand, they can internalize the tax by lowering their prices. The result is that consumption stays constant but producer profits decrease. That might work with a modest tax, but for an industry already struggling to raise pigs and cattle worth less than their production cost, and now faced with the loss of subsidies, internalizing a 50 percent tax would be impossible. Further, even if producers found a way to internalize a significant part of the Meat Tax—say, lowering prices by 10 percent, the tax and other proposed changes would still achieve most of their intended effects (including generating a projected $20 billion cash surplus for the US Treasury).

It's too high. Most states impose a sales tax of 5 to 8 percent—why not start with something in that range?

That wouldn't work for the tax's main goal: to discourage consumption. A single-digit tax would simply legitimize current consumption levels and the damage they cause, and it would do little to lower those levels. Further, a tax that low might let producers internalize all or most of the price increase by lowering their prices, which would further weaken the tax's effect.

The tax will reduce consumer spending, lower economic activity, and hurt GDP.

The tax doesn't seek to reduce overall spending, just to shift some spending from animal foods to other foods. It's not a diet, and there's no reason consumers won't continue to devour the same number of calories they did previously—albeit from healthier sources—which means buying substitute foods. Further, the tax credit to individual taxpayers provides a cash boost and a significant source of stimulus funding. So there's no reason why the tax should cause an overall reduction in spending.

It will put animal food producers out of business and eliminate jobs.

It's true that a Meat Tax will cause some job loss in animal agriculture. But there are several mitigating factors. For one thing, people will continue to eat—just different foods. The size of the employment pie doesn't need to change, even if it gets sliced up differently. In other words, the jobs sacrificed in meat, fish, eggs, and dairy will be offset by new ones in sectors that raise crops for humans (rather than for livestock) or that manufacture plant-based foods. These sectors will also need additional inputs like seeds, equipment, and transportation, all of which will see job growth or replacement as a result.

In fact, if you had a job in animal food production, you might be *glad* to change it. The US Bureau of Labor Statistics reports that personnel in slaughterhouses and milk production are more than twice as likely as the rest of the working population to suffer a workplace injury.[43] A *New York Times* article found Smithfield pork slaughterhouse workers performing painful and repetitive tasks for $10.24 an hour (in current dollars).[44] Further, racism is rampant in slaughterhouses, not only in self-segregating cafeterias and locker rooms, but in the way the work is doled out: whites get the best jobs (supervising or mechanic work), followed by Native Americans (warehouse work), then blacks (kill floor), with Mexicans (butchering) at the bottom.[45]

But for workers in the animal food system, injuries and racism are just the beginning. In their 2010 book *Food Justice*, Robert Gottlieb and Anupama Joshi detail a slew of injustices faced by laborers in American agricultural sectors like fishing and meat processing.[46] Ninety percent of agriculture workers lack health insurance. One-third are undocumented. One-quarter live in conditions that one state agency called "extremely overcrowded." As a result of such odious work conditions, the Smithfield plant profiled in the *New York Times* article experienced 100 percent annual turnover; each year, five thousand workers quit and five thousand new workers are hired.

It's true that a Meat Tax will put some animal food producers out of business. But to some extent, particularly for factory farm operators, isn't that the goal? We've seen that the problems of meatonomics are supply driven, or spurred by producers. These animal agribusinesses are responsible for offloading more costs onto American society than any other industry. We've seen that animal food producers routinely seek to deceive us about their products' health effects and their treatment of animals. Must Americans continue to prop up factory farmers and fishers at the expense of so much that's wrong with their industries, merely because of the argument that they help the US economy? In fact, we've seen in this chapter that eliminating half of their business will actually *help* the economy in a big way.

Despite their protestations, disfavored industries are regularly targeted by social movements that seek to curb or eliminate them. When Britain sought to ban the slave trade in the 19th century, staunch opposition came from the city whose ships carried more than half of Europe's slaves—Liverpool. Liverpudlians wrote a song that began, "If our slave trade be gone, there's an end to our lives, Beggars all we must be, our children and wives."[47] Similar sentiments were expressed in the United States, where Thomas Dew, the future president of the College of William and Mary, wrote in 1832, "It is in truth the slave labor in Virginia which gives value to her soil and her habitations—take away this and . . . the Old Dominion will be a 'vast howling wilderness'—the grass shall be seen growing in the streets, and the foxes peeping from their holes."[48] Unfortunately, sometimes social and economic progress means temporary discomfort for a few.

It's overly paternalistic.

Americans like the right to do as we please, and we don't want government telling us what to eat. Is the Meat Tax too paternalistic? To look at it another way, does the tax diminish individual choice in much the same way that the existing system does? Consider one of the simplest definitions of, and objections to, paternalism, by the philosopher John Stuart Mill:

> The only purpose for which power can be rightfully exercised over any member of a civilized community, against his will, is to prevent harm to others. His own good, either physical or moral, is not a sufficient warrant. He cannot rightfully be compelled to do or forbear because it will be better for him to do so, because it will make him happier, because, in the opinions of others, to do so would be wise, or even right.[49]

Certainly, on one level, the Meat Tax does seek to change people's behavior for their own good. But on another, and perhaps more important level, the tax seeks to prevent harm to others as Mill contemplated. Some of the others routinely harmed by American animal food production practices, whom this proposal would in some measure protect, include:

- Farmers in developing nations who are hurt by our subsidies and dumping practices.

- Residents of Asia and South America, where most of our fish is caught or grown, whose local ecosystems are routinely compromised for our dining pleasure.

- Future generations of Americans who must someday pay the deferred, externalized costs of producing animal foods today.

- 16 million Americans who consume little or no animal foods yet still incur the industry's massive externalized costs.

- 60 billion land and marine animals killed yearly to feed Americans, and billions more marine animals killed as bycatch— including millions of endangered animals like albatross and sea turtles.

- Millions of US bee colonies destroyed by pesticides and reduced foraging opportunities resulting from monocrop planting— most of which involves feed crops.

- Thousands of environmental features like lakes and rivers polluted by animal waste, which some scholars believe have, or should have, independent legal rights.[50]

Yes, the Meat Tax is a little paternalistic. But in light of the many others it seeks to help, not overly so. Besides, Americans have a history of imposing—and tolerating—paternalistic sin taxes on items like alcohol and tobacco. There's little reason to view the Meat Tax differently.

It's not politically feasible.

This might be the most daunting criticism. No one likes a new tax. And because this proposal targets one of the most powerful and well-funded industrial sectors in the country, it's like David against an army of Goliaths.

But you have to start somewhere, and as Gandhi advised while dismantling British rule in India, "If you don't ask, you don't get." Pundits once thought it inconceivable that slavery would be abolished, that Congress would pass the Civil Rights Act, or that Big Tobacco would be held legally accountable for destroying the lungs and lives of millions of Americans. A decade ago, few would have thought we'd soon have a black president—and certainly not for two terms. Change *does* happen, often in swift and surprising ways.

Moreover, I optimistically submit that the Meat Tax *is* feasible. Tax credits will put cash in most voters' pockets and help stimulate the economy—ideas that are always popular. The credits also ensure that the typical consumer's disposable income available for food will remain constant. Further, the tax is lower than the average tobacco tax of 72 percent, which lawmakers have increased steadily over the years and are likely to raise even more in the future. The Meat Tax's many social benefits—like saving lives, restoring land, protecting fisheries, making water cleaner, reducing animal suffering, and helping reduce global warming—are big selling features. Another boon is the huge cash surplus it generates each year, which can be used to fund social programs and reduce individual income taxes.

If you don't ask, you don't get. The institutional changes proposed in this book require help from lawmakers, which requires voters to speak loudly and clearly. Here's one idea. Ask your state and federal legislators to end subsidies to animal foods and to start taxing those foods. Their contact information is here: *usa.gov/contact/elected.shtml*. As cultural anthropologist Margaret Mead said, "Never doubt that a small group of thoughtful, committed citizens can change the world. Indeed, it is the only thing that ever has." Make a phone call, send a letter or an email, compose a tweet or go on Facebook, or better yet, meet your lawmaker in person. Although we've seen lawmakers often side with corporate interests, a survey of congressional staff found that the best way to influence lawmakers is not through paid lobbyists but through "in-person visits by constituents."[51] So anyone really *can* make a difference.

Getting and Spending

Two centuries ago, when the Industrial Revolution was still in its infancy, poet William Wordsworth observed of modern, workaday life: "Getting and spending, we lay waste our powers." I can relate. I've worked off and on as a business lawyer for more than two decades, which means I've spent a lot of time putting cash in the pockets of clients like Frank. As you may know from personal experience, money-driven work like that is largely devoid of social significance other than to the few people who benefit from it. It drains your passion and leaves you little time or energy to do much else.

But a few years ago, working pro bono for a nonprofit organization, I discovered how it felt to put my powers to better use. After sending a letter and engaging in a minor negotiation, I convinced authorities to allow activists protesting animal cruelty to enter a shopping mall. I was thrilled by the tiny win (although admittedly, I kept my day job). Describing my feelings later to activist Dina Kourda, I said, "I've been practicing law for years, and this is the first time . . ." My voice trailed off, but Dina finished my sentence, hitting the nail on the head: ". . . you did something that actually mattered?"

It's easy to get caught up in the day-to-day getting and spending that life demands. Sometimes it just takes a single, jarring event—a blog, a movie, a book—to remind us that some things in life matter more than money. And the next time that particular realization grabs you like a warm embrace, there's plenty you can do without giving up your day job. Consider the words of Gandhi: "You may never know what results come from your action. But if you do nothing, there will be no result."

Food for Thought

- We can act now to reverse the social and financial costs of meatonomics. Individually, each of us can help by consuming less meat, dairy, fish, and eggs.

- Collectively, we can address these problems with a Meat Tax. The tax-driven policy changes would save 172,000 human lives and 26 billion animal lives, generate a $32 billion annual cash surplus, permit the return of one-sixth of the contiguous United States to native habitat, and have the same effect on carbon emissions as garaging all American highway vehicles, boats, and ships.

- The best part is, by making price levels more accurate and decision making more meaningful, these changes will empower consumers to make better-informed choices.

ACKNOWLEDGMENTS

This book sometimes asks people to think about animals and our relationship to them, and I'd like to remember two of the many animals who got me interested in this particular subject. In 2004, a law enforcement officer shot to death a cougar sleeping in a tree in Palo Alto, California. With the state's populations of humans and cougars then at 35 million and four thousand, respectively, this incident led me to reflect on the economic principle—inconsistently applied, it turns out—that value is based on scarcity.

And a few years later, the slaughterhouse video *Glass Walls* showed a cow dangling upside-down from one hoof in a fast-moving butchering machine. Already past the stunning station (without being stunned) and headed for the belly-ripping station, she continued to bellow and struggle to free herself. We don't often see animals as individuals, but when we do, it can leave a lasting impression. These individual animals, and many others whose lives or deaths I've seen— sometimes just in brief glimpses—have changed the focus of my own existence.

I am indebted to the many people who helped with this book— particularly those who provided comments on the manuscript simply because I asked. As the combined result of their good-natured generosity and my shameless begging, I had the tremendous good fortune to be advised in this effort by more than thirty people. Seven readers have advanced degrees in economics. Ten have PhDs. Twelve are lawyers. Some fit more than one category. While this feedback dramatically improved the book's accuracy and objectivity, I remain solely responsible for any errors. Further, I note that several of those who commented disagree, in varying levels, with my analysis and/or recommendations.

I am particularly grateful to Donald Garlit, Claire Kim, John Maher, and Michael Pease for heroically spending the many hours necessary to read and comment on the entire manuscript. For their valuable help in reviewing parts of the manuscript, I also owe deep thanks to John Boik, Chris Bryan, Karen Davis, Carol Glasser, Michael Harrington, Chris Holbein, Julie Jaffe, Miles Jaffe, Melanie Joy, Dina Kourda, Tom Lillehof, Dara Lovitz, Tania Marie, F. Bailey Norwood, Robert Ranucci, Kendra Sagoff, Mark Sagoff, Larry Simon, Max Simon, Michele Simon, Janice Stanger, Paul Wazzan, James McWilliams, and Bill Weissinger. And special thanks to Erin Evans for her valuable research and editorial assistance.

A number of veterans of the book trade were key to this work's publication, and I am grateful for their expert assistance. My agent, Lindsay Edgecombe, provided invaluable advice at each step, guiding me through the publishing process with energy and savvy. My editor, Caroline Pincus, backed this book with courage, enthusiasm, and foresight. Josh Chetwynd provided terrific editorial help, excising redundancies and lawyer-speak with the skill of a surgeon. Thanks also to Ali McCart for her astute and meticulous copyediting, Vanessa Ta for her excellent production editing, and Bonni Hamilton and Anne Sullivan for their zealous and creative promotional efforts.

I'm especially grateful to my partner, Tania Marie, for her extraordinary patience, support, and encouragement. The past few years saw many a sunny Saturday or Sunday morning spent in the house, working. She brightened those indoor hours with her sparkle.

Finally, I must also recognize Joy, Gaia, Boojum, and Sweet Pea—the rabbit, tortoise, and two cats who share our home and whose visits punctuated and enlivened many a writing session. The cats even gave their own input, walking on the keyboard every so often to insert random strings of characters in the manuscript. These individual animals' unique personalities and behaviors provide a daily reminder that they, like all sentient beings, live to pursue their own versions of happiness.

Animal Foods and Human Health

Don't we need milk for healthy teeth and bones? Aren't fish a great source of omega-3s? Isn't an egg a day good for you? Isn't meat a necessary source of iron, Appendix B$_{12}$, and other nutrients? As musician Steve Albini cautions, "Doubt the conventional wisdom unless you can verify it with reason and experiment." This appendix explores a number of widely held beliefs related to the health effects of animal food consumption. Specifically, the first section addresses issues in beef and pork, the second looks at dairy, the third eggs, and the fourth fish. As we'll see, reason and experiment don't always support popular assumptions.

A Red Flag for Red Meat: Issues with Beef and Pork

Consider a common belief about animal protein: that its quality is better than plant protein's. Along those lines, beef and pork producers routinely trumpet the high-quality protein their products supposedly deliver.[1] Where do these groups get the idea that animal protein is high quality and plant protein is low quality?

This notion seems based on the idea that some proteins contain a more complete set of the eight essential amino acids than others. Amino acids are building blocks of protein that our bodies cannot produce independently. Despite research to the contrary, the idea developed that only animal foods contain all essential amino acids, and that has led to these foodstuffs being called, somewhat hyperbolically, "complete proteins." Yet plants can also provide a complete set of essential amino acids. Writing of the common misconception that plants lack certain essential amino acids, physician and nutritionist John McDougall said, "Any single one or combination of . . . plant foods provides amino acid intakes in excess of the recommended requirements," and for a vegetarian who consumes sufficient calories, "it is impossible to design an amino acid–deficient diet."[2] It's unclear why the myth persists that plants lack amino acids and are incomplete. As early as 1966, a clinical study of amino acid intake in meat-eaters, vegetarians, and vegans found that "each group exceeded twice its requirement for every essential amino acid and surpassed this amount by large amounts for most of them."[3]

Paradoxically, despite the misdirected focus that consumer messaging often places on animal foods providing more essential amino acids than plants, animals actually have no independent ability to create amino acids. Only plants and bacteria can generate protein's building blocks. As nutritionist Janice Stanger notes, "Animal protein is recycled plant protein."[4] When humans obtain protein from vegetables, we simply cut out the animal middleman.

Meat and Human Evolution

In the animal kingdom, herbivores get all the protein they need for healthy muscle and tissue development from plants. Moreover, power, speed, and body size often favor those who eat plant food, not animal food. Thus, nature's strict herbivores include large animals like elephants, rhinoceroses, giraffes, and fast animals like gazelles, antelopes, and horses. Not only do these vegans get plenty of protein from plants, but their robust physiques and longevity belie the feared trade-off that humans typically associate with eating plants.

But humans are different from plant-eating animals, some say, because our bodies have *evolved* to eat meat. This argument typically advances in three parts: human anatomical features, such as canine teeth, are similar to carnivores'; vegetarian diets don't provide necessary nutrients like protein, calcium, iron, and vitamin B_{12}; and hunting and meat eating have been natural human or hominid occupations for millions of years. In fact, a closer look at these issues tends to point toward the opposite conclusion: it seems humans, just like gorillas, evolved to eat plants.

In a study examining the comparative anatomy of carnivores, omnivores, herbivores, and humans, physician Milton Mills compared nineteen anatomical features from the four groups. Mills found that humans most closely resemble herbivores, not carnivores or omnivores, in all anatomical features related to eating.[5] Thus, like herbivores but unlike carnivores or omnivores: our saliva contains enzymes to digest carbohydrates; our intestines are long, not short; our mouth opening is small, not large; our stomach's pH is 4 to 5, not 1; we chew food rather than swallow it whole; we have flattened nails instead of sharp claws; our molars are flattened, not sharp; our incisors are broad and flat, not short and pointed, and our canines are short and blunted, not long and sharp. These features all support plant consumption and suggest that in humans, the evolutionary process selected features associated with eating vegetation. For instance, a long intestinal tract and a higher stomach pH are appropriate to digest plant material, not flesh; and flat molars, blunted canines, and a predisposition to chew food distinguish us from carnivores and omnivores who typically rip and swallow food whole. In the anatomy of eating, it seems we have much more in common with Bambi and Bullwinkle than with Lassie or Leo.

The argument that humans cannot obtain adequate nutrients from a vegetarian diet is a common misconception. In fact, as seen in chapter 2, protein is available in every kind of plant food. Iron and calcium are readily available in a wide variety of plants. Moreover, in each instance, the plant sources of these nutrients do not promote disease, while the animal sources promote illnesses like cancer, heart disease, and diabetes.

Vitamin B_{12} is one nutrient some vegans might have a problem obtaining from plants, but not because it's not present naturally. In fact, neither plants nor animals are capable of independently producing B_{12}; the vitamin is produced only by bacteria.[6] These bacteria are typically present in unwashed vegetables, but not in the triple-washed, hermetically packaged vegetables most people eat today. Thus, while it's appropriate for strict vegetarians to take B_{12} supplements unless they grow and fertilize their own crops, this is merely the result of life in industrialized society and not a feature of body design. Furthermore, our bodies need only the tiniest amount of B_{12}. Those who eat lots of meat may get too much, and high levels of B_{12} are associated with a higher risk of prostate cancer.[7]

As to the argument that hominids have been eating meat for aeons, the archaeological record is less than conclusive. One school of thought posits that early humans were predominantly scavengers, not hunters, which explains the presence of animal bones at early human sites.[8] The theory that meat consumption in early humans was merely opportunistic, rather than the product of complex social activities like group hunting, tends to counter the argument that human evolution has been closely related to dietary changes like increases in animal protein consumption.

Another interesting line of research postulates that early humans were mainly prey, not predators, engaging in social behaviors like group living for self-defense rather than hunting. According to archaeologist Robert Sussman, our hominid ancestors were vegetarians who, because of body morphology, "simply couldn't eat meat."[9] These early humans inhabited a hostile landscape populated by ten times more predators than today, and one in ten hominids became another animal's dinner. Sussman and others argue that the concept of early humans as hunters has been radically overstated; in fact, evidence suggests that systematic hunting did not begin until relatively recently within the total scope of human evolution.[10]

While it is undeniable that humans have hunted for millennia, it does not follow that our bodies evolved to eat meat. We're opportunistic, intelligent, and highly capable of developing and using technology. To look at it another way, we readily build and use airplanes, although our bodies did not evolve to fly.

Dairy Dairy, Quite Contrary: Issues with Milk, Cheese, and Butter

Through aggressive government marketing and artificially low prices, meatonomics encourages Americans to consume huge quantities of dairy. Each American takes in about 2 pounds of dairy every day. That's nearly three times the worldwide average and, according to experts, considerably more than our bodies can safely process.[11] As with meat, these high consumption levels damage our health and cost billions of dollars to treat. A brief overview of dairy's health effects shows why.

Mammals have mammaries. That makes us unique in the animal kingdom, both in producing milk and in drinking it when young. Milk is nature's way of promoting rapid growth and boosting the immune system of infants, which is why milk is stocked with antibodies and nutrients like protein, calcium, and vitamin C. For human babies unable to chew or digest solid food, milk is a great delivery vehicle for these nutrients. For adults and children past weaning age, however, a sizable body of research suggests otherwise. These studies find that when fed to non-infants, especially at the levels Americans consume it, dairy is not only unnecessary but harmful.¶ As Dr. Michael Klaper said, "The human body has no more need for cows' milk than it does for dogs' milk, horses' milk, or giraffes' milk."[12]

Consider protein, the building block of muscle and tissue development. Because natural selection among prey animals like cattle favors those whose young grow quickly, bovine infant formula, or cow's milk, has triple the protein content of human's milk. This high protein content helps calves gain 2 pounds a day during the first nine months of their lives. It also helps human children who drink lots of cow's milk grow faster than those who drink less.[13]

This rapid pace of growth might appeal to parents who associate fast growth with good health. However, clinical studies question this need for speed, finding that children who grow quickly are more likely than others to develop cancer later in life.[14] Do human children really need to grow as fast as cattle? Maybe not. For one thing, unlike prey animals, human babies can afford to grow slowly because their parents protect them from predators.

Cancer

Dairy promotes rapid cell growth, but this is a double-edged sword. One problem with this process is it spurs the development of both

---

¶ Milk, the source of all dairy, is the main ingredient in cheese, butter, cream, ice cream, yogurt, and whey.

healthy and unhealthy cells. As a result, cancer cells develop in ways that evade or overwhelm the body's natural capacity to attack and kill them. The biggest culprit in this area appears to be the protein casein, one of the main ingredients in milk. According to *The China Study* coauthor T. Colin Campbell, casein is an "exceptionally potent cancer promoter."[15] This is *not* just a problem for milk or cheese gluttons. In fact, at levels well below the two pounds consumed by the typical American each day, research consistently finds dairy causes cancer.

Take prostate cancer, a disease which at least sixteen clinical studies link to dairy consumption.[16] In one study, men who drank more than two glasses of milk daily were found to have a significantly greater risk of prostate cancer than those who drank no milk.[17] The daily danger threshold of two glasses is *one less* than the USDA recommends and almost *two less* than the American average.

Among women, research finds that dairy—unfortunately, an equal opportunist—can cause both breast and ovarian cancer.[18] In the latter case, two large cohort studies find that women who consume just two servings of dairy daily—again, *one less* than recommended and *two less* than the average—have a significantly higher risk of ovarian cancer than those who consume less.[19] There's a pattern here: at consumption levels well below those the USDA recommends or Americans practice, dairy promotes disease.

Bone Health

The most surprising news about dairy may be its effect on bone density and health. We know that calcium is important to preserve bone density and prevent osteoporosis and that milk is an abundant source of calcium. But milk's acidic properties can affect our ability to fully utilize its calcium, and this can lead to some odd results. For example, a review of fifty-eight clinical studies evaluating the importance of dairy consumption to healthy bone growth in children and adolescents found no significant link between milk consumption—or calcium consumption—and healthy bone development.[20] The authors of this study conclude:

Neither increased consumption of dairy products, specifically, nor total dietary calcium consumption, has shown even a modestly consistent benefit for child or young adult bone health. Conclusion: Scant evidence supports nutrition guidelines focused specifically on increasing milk or other dairy product intake for promoting child and adolescent bone mineralization.[21]

Among older populations, research has yielded even weirder findings about dairy and bone health. A Harvard study found that women who drank two or more glasses of milk daily had a significantly greater risk of hip fractures than those who drank one glass or less weekly.[22] And a study of elderly Australians found that those who consumed the most dairy had twice the risk of hip fracture compared to those who consumed the least.[23] Coincidentally, as with prostate and ovarian cancer risk, the critical level at which dairy is risky for bone health—two servings a day—is below both the three recommended by the USDA and the nearly four servings (in milk or equivalents) consumed by Americans daily.

There are two reasons why consuming dairy can lead to problems with bone density like osteoporosis. First, dairy's acidic pH causes the body to release calcium (an alkali) from bone to counter the acid's effects and restore healthy pH.[24] Second, high levels of calcium consumption over long periods of time impair the body's ability to regulate production of the hormone calcitriol, which controls calcium absorption and excretion.[25] Inappropriate calcitriol production can lead to excessive release of bone cells and reduction in bone mineral density.

The body tends to adjust to higher and higher levels of calcium intake, that is, to raise its point of balance to equal the level of calcium consumption. This compensation mechanism means those who consume calcium supplements can suffer the same bone density problems as those who consume large quantities of dairy.[26] One of the leading scientists in this field, the late Harvard nutritionist D. Mark Hegsted, wrote of the irony of this phenomenon:

As we encourage people to consume more calcium, their calcium requirement—as defined by calcium balance—increases, with no end in sight. Numerous population studies in the U.S. and elsewhere have analyzed the relationship between dietary calcium intakes and fractures, finding that surprisingly, high calcium intakes do not protect against fractures.[27]

On the contrary, high intakes of calcium *lead* to high fracture rates. In Hong Kong, daily calcium consumption is less than 500 milligrams per day, and females suffer hip fractures at the rate of thirty-five cases per hundred thousand people. Americans consume at least twice as much calcium, averaging about 1,000 milligrams per day. Yet our per capita incidence of female hip fractures is three times higher than Hong Kong's.[28] This research is consistent with other studies that find bone fractures are more common in Western countries, where dairy consumption is high, than in countries where little dairy is consumed.[29]

The studies on calcium and hip fractures are not exactly breaking news—they were published in the 1980s and '90s. Moreover, modern research suggests that people seeking to improve bone density should exercise more and eat plant-based calcium sources like kale and spinach (a plan that seems to work for herbivores like horses and elephants). Why then do Americans still turn, somewhat futilely, to dairy and calcium supplements for bone health? We do so because despite the high correlation between dairy consumption, calcium intake, and hip fractures, the dairy industry and the USDA continue to recommend that Americans aged nineteen to fifty consume 1,000 milligrams of calcium per day. This recommendation appears to have one goal: to sell dairy products.

Kids and Milk

For many, it will be astounding to learn that kids don't need cow's milk and would actually be better off without it. Why is this news so astonishing? Because just as we know the sun rises in the east, we know children *need milk* to grow up big and strong. Yet one of the pieces of information to emerge from recent studies is that kids grow up healthier when they *don't* consume significant amounts of dairy (other than their own mothers' milk) or other animal protein. Certainly, children need their own mothers' breast milk while in infancy. But otherwise, and after the age of weaning, there is little in the clinical literature to suggest that children need cow's milk or other animal foods for healthy development.

In fact, studies show the opposite. Girls who develop too quickly, for example—a phenomenon associated with drinking cow's milk—are more likely to develop breast cancer later in life.[30] In one study of British girls, researchers found fast growth was associated with more than a 50 percent greater risk of breast cancer, leading the authors to conclude that "women who grow faster in childhood and reach an adult height above the average for their menarche [first menstruation] category are at particularly increased risk of breast cancer."[31] This heightened cancer risk may stem from the fact that other than in infancy, mammals' bodies are simply not designed to drink milk. After humans are weaned, we stop secreting both the enzyme rennin that breaks down milk protein and the enzyme lactase that digests lactose (milk sugar) and transforms it into sugars that our body can use (glucose and galactose).

Doctor Benjamin Spock wrote the bestselling book *Baby and Child Care* and, a decade after his death, remains one of the most influential pediatricians on the planet. Spock staunchly opposed feeding cow's milk or other animal foods to children of any age. In the "Foods to Avoid" section of his book, Spock wrote:

> *Meat:* Children who grow up without developing a taste for meat, poultry, or fish will carry a tremendous advantage through life. Their tastes will not be oriented toward these products, all of which contain fat and cholesterol that can contribute to weight problems, heart disease and some cancers. They will also be at much less risk of infection with the bacteria that often taint meat products.
>
> *Dairy products:* Nondairy milk, particularly soy milk has real advantages over cow's milk and other dairy products. These

products are free of animal fat, animal protein, and lactose sugar, while still providing excellent nutrition.[32]

A number of children's doctors agree, including Dr. Frank A. Oski. While serving as director of pediatrics at Johns Hopkins University, Oski said, "There's no reason to drink cow's milk at any time in your life. It was designed for calves, not humans, and we should all stop drinking it today."[33]

Egg-cess: Issues in Eating Eggs and Chicken

Are eggs really the "perfect protein"? They do provide protein and other nutrients. But they're also an abundant source of cholesterol; a single large egg contains 212 milligrams of the stuff. Because ingested cholesterol is never necessary but always harmful to our bodies, that's 212 milligrams more than we need and two-thirds more than the USDA's liberal, recommended daily maximum of 300 milligrams.[34] The Jekyll-and-Hyde nutritional personalities of eggs seem to fit a pattern familiar in animal foods—they contain some helpful nutrients, but they also include plenty of disease-promoting substances. In the case of eggs, research has created controversy in the scientific community over whether their nutritional benefits outweigh the risks associated with their high cholesterol content.

Many studies show that egg consumption increases blood cholesterol, a phenomenon linked to increased risk of heart disease.[35] On the other hand, a well-publicized 2007 paper by Adnan Qureshi and colleagues found that adults who eat more than six eggs per week have no greater risk of stroke or heart disease than those who eat no eggs.[36] While some point to the first set of studies to argue that eggs cause heart disease, others—notably, the egg industry—cite the Qureshi article as evidence of eggs' harmlessness. Because this study's design and methodology are similar to those of other inquiries finding that animal foods are healthy, such as the Siri-Tarino piece discussed in chapter 1, it merits a closer look.

Using the cohort research model, Qureshi's team compared several populations of egg eaters to determine their incidence of disease over time. Qureshi's team adjusted for confounding factors like age, gender, race, and cigarette smoking. But just as the Siri-Tarino researchers did, the Qureshi team failed to adjust for the most important confounding factor: regular consumption of other animal foods. American males over the age of twelve routinely consume more cholesterol than recommended, in some cases by 100 milligrams or more daily.[37] Since many Americans are *already* in a high-cholesterol category, which leads to heart disease for one in three, it's easy to see why adding an egg a day might not materially increase this risk.

In fact, in order to make their analysis representative of the US population, the Qureshi team *chose* cohorts whose members were overweight (with BMI over 25) *and* had blood cholesterol levels of 220 mg/dl or higher. That's 10 percent above the recognized safe cholesterol level of 200 mg/dl and 50 percent higher than the average of 146 mg/dl among vegans.[38] In other words, this study looked at whether overweight people with high cholesterol, already substantially at risk for heart disease, would get any worse if they ate an egg a day. That's like studying the risk of cigar smoking among people who already smoke two packs of cigarettes a day.

Dietary cholesterol is present in animal foods but not in plant foods. So the only accurate way to measure the cholesterol-related effects of eating eggs is to compare a cohort of vegans with one whose only source of dietary cholesterol is eggs. Since the Qureshi study failed to control for the confounding effects of cholesterol intake, it's little wonder the study found no difference in disease risk between groups.

On the other hand, researchers *do* find a link between egg consumption and disease when they compare cohorts who eat less animal foods than Americans do and for whom a single egg is thus a more significant part of their diet. One study involving Japanese cohorts noted health benefits associated with "limiting egg consumption."[39] These researchers found links among egg consumption, blood cholesterol levels, and mortality rates among women "in geographic areas where egg consumption makes a relatively large contribution to total dietary cholesterol intake."[40] Plainly stated, the harmful effects of eggs are more clearly evident among those who eat less animal foods to begin with.

Despite the Qureshi study's design flaws, it's a favorite among egg industry marketers. The American Egg Board receives $21 million in yearly federal-mandated checkoff money.[41] These funds power the website *eggnutritioncenter.org*, where the Qureshi study gives credibility to misleading messages like "An Egg a Day is MORE Than Okay!"[42]

Not to get too metaphysical, but is it possible to discuss eggs without mentioning chicken? Actually, chicken-related illnesses like cancer, heart disease, and food poisoning are discussed in the main text. As shown in chapter 6, there's little that's healthy about chicken.

Something's Fishy: Issues in Eating Seafood

Like eggs, fish are a nutritional paradox. On one hand, they're a good source of omega-3 fatty acids, which can improve infant brain development and reduce the risk of coronary heart disease.[43] On the other hand, almost without exception, fish contain mercury, a neurotoxin that can damage the nervous system and cause cognitive disabilities.[44] And they frequently contain polychlorinated biphenyls (PCBs), which cause cancer and, in infants, neurological and motor control problems.[45] Like all animal foods, fish also invariably contain cholesterol (in some cases at levels as high as ground beef—see table 6.1 in chapter 6), and this substance causes heart disease. Finally, because *all* animal protein—including fish—promotes cancer, a diet high in seafood increases one's risk of cancer relative to a plant-based diet.[46]

Where does all this mercury and PCB come from? The mercury seems to come mainly from coal-fired power plants. PCBs, on the other hand, were used in manufacturing in the United States until they were banned in 1977, but they've lingered in our water and soil for decades and, as a result, persist in the environment. Both PCBs and mercury accumulate in the fatty tissues of fish and other animals, which means they are passed up the food chain until eaten by humans.

In a seven-year study of fish in 291 US waterways, the US Geological Survey found that *every one* of more than one thousand fish it examined contained some mercury. One-quarter of the fish contained mercury at levels unhealthy for human consumption, and two-thirds were contaminated at levels unhealthy for consumption by other mammals.[47] It's not just fish in inland waterways that are toxic with mercury; high levels of mercury are routinely found in deep-sea fish as well.[48] And PCBs are equally prevalent: another study found that thirty-nine of forty fish sampled contained PCBs in measurable quantities.[49]

The nutritional paradox surrounding fish leads to complicated advice on whether and when to eat it, dispensed in articles with titles like "Fish Is Good—Fish Is Bad" and "Americans Confused about Health Effects of Eating Fish."[50] In one particularly bizarre piece in the *Journal of the American Medical Association,* the authors note that omega-3s may help new or expectant mothers, but a few paragraphs later, warn that new or expectant mothers "should avoid four types of fish that are higher in mercury content."[51] Other articles contain charts advising which fish have acceptable levels of mercury and PCBs and how often one can eat them.[52]

If this seems like a lot of risk and effort to obtain omega-3s, consider a non-fish, nontoxic source for your beneficial fatty acids. Omega-3s are readily available in flax, hemp, soy, and walnuts. You won't need a chart to remember which you can eat and when, and none of these plant-based sources causes cancer, neurological defects, or developmental impairment.

Appendix B

Summary of the Annual Externalized Costs of US Animal Food Production (in billions)[1]

	REFERENCE YEAR	NOMINAL COST	2012 INFLATION-ADJUSTED
Health Care			
Heart disease	2008	$133.3	$143.1
Cancer	2007	75.6	84.3
Diabetes	2007	55.1	61.3
Antibiotic resistance	2000	17.5	23.6
Salmonella poisoning	2010	1.3	1.4
E. coli poisoning	2010	0.3	0.3
		$283.1	$314.0
Subsidies			
Federal loans, insurance, research, etc.	2012	$18.2	$18.2
State irrigation subsidies	2009	13.5	14.6
Other state and local subsidies	2009	2.0	2.1
Federal irrigation subsidies	2009	1.1	1.2
Fishing subsidies	2003	1.8	2.3
		$36.6	$38.4
Environmental Costs			
Soil erosion	1992	$9.4	$15.4
Climate change mitigation	2011	9.2	9.4
Pesticide use on feed crops	2003	5.0	6.3
Real property devaluation	1999	1.8	2.5
Eutrophication of waterways	2008	1.1	1.2

	REFERENCE YEAR	NOMINAL COST	2012 INFLATION-ADJUSTED
Manure spreading vs. storage	2005	1.2	1.4
Manure lagoon repair	2003	0.8	1.0
		$28.5	$37.2
Cruelty	2011	$20.3	$20.7
Fish Production			
Overfishing	2010	$1.4	$1.4
Bycatch	2010	2.3	2.4
Fish farming	2009	0.6	0.7
		$4.3	$4.5
Total		$372.8	$414.8

Economic Effects of Proposed Meat Tax and Support Changes

This appendix shows calculations for the main economic effects expected to result from this book's proposed changes to federal tax and support policy. These outcomes include: (1) reductions in subsidies and other support to the animal food industry, (2) a drop in consumption of animal foods, (3) a new tax burden for taxpayers, (4) a new tax credit for taxpayers, (5) an increase in tax revenue, (6) higher federal government outflows associated with the tax credit, (7) lower state and federal government outflows under Medicare and Medicaid programs, (8) a cash surplus for the federal government, and (9) decreases in externalized costs. As a reminder, economics is not a precise science, so while these figures may seem meticulously calculated, they are just estimates. You don't really know how something will play out until you try it, although you can certainly make predictions.

Weighted Average Elasticity Rate

The following calculations use a constant of 0.65 as the price elasticity of demand for animal foods. This constant is a weighted average based on the individual fair market values of animal foods produced yearly and their respective price elasticities, as shown in table C1. Note that "production value," roughly analogous to wholesale value, is used rather than retail value because retail sales data for individual animal foods is not uniformly available.

TABLE C1 Animal Food Production Values, Elasticities, and Weighted Average Elasticity Rate (dollar amounts in billions)

ITEM	PRODUCTION VALUE[1] (P)	MEAN PRICE ELASTICITY[2] (E)	P X E	WEIGHTED AVERAGE ELASTICITY RATE ((P X E)/P)
Beef	$36.70	0.75	27.53	
Pork	16.10	0.72	11.59	
Chicken	23.70	0.68	16.12	
Turkey	4.40	0.68	2.99	
Dairy	31.50	0.65	20.48	
Fish	23.30	0.50	11.65	
Eggs	6.50	0.27	1.76	
	$142.20		92.12	0.65

Changes in Subsidies and Other Support

The proposed support changes affect two categories. First, there's the gross figure of $28.9 billion that the USDA spends supporting US farmers with loans, insurance, research, marketing assistance, and other help (the research and marketing subsidy).[3] My recommendation is to eliminate the portion of this subsidy related to animal food production, which is roughly $18.2 billion.[4] Loss of this subsidy will have an effect on both producers and consumers.

From the producers' perspective, because the Meat Tax will cause consumption (and production) of animal foods to decline by 32.5 percent (see "Drop in Consumption" below), the portion of this subsidy on which producers would rely in order to maintain the status quo after adjusting sales for the effects of the Meat Tax is 67.5 percent (1 – 0.325) of $18.2 billion, or $12.3 billion. In other words, following the drop in consumption, a subsidy of only $12.3 billion will allow producers to maintain production at the new, lower level.

In the absence of data predicting how producers are likely to respond to the loss of certain services, it is reasonable to assume that they will choose to discontinue roughly half of these formerly subsidized services and continue half at their own expense ($6.1 billion). And in light of data showing that producers typically internalize some portion of increases in production costs (ranging from a minority to a majority of such costs), it is reasonable to assume further that producers will internalize roughly half of these costs and pass the other half ($3 billion) on to consumers.[5] The resulting price increase is roughly 1.8 percent of the total retail sales of animal foods, after reducing this sales figure for the effects of the Meat Tax (see "Drop in Consumption" below). Table C2 shows the effect of applying the demand elasticity constant of 0.65 to this increase: eliminating this subsidy will cause consumer demand to decline by about 1.2 percent.

TABLE C2 Effects of Eliminating Research and Marketing Subsidy (dollar amounts in billions)

TOTAL SUBSIDY	$28.9
Percent related to meat and dairy	63%
Net subsidy related to meat and dairy	$18.2
Percent to maintain status quo under Meat Tax	67.5%
Net subsidy denied to industry	$12.3
Percent internalized by industry	50%
Net costs internalized by industry	$6.1
Percent passed through to consumers	50%
Net costs passed through to consumers	$3.1
RETAIL SALES OF ANIMAL FOODS UNDER MEAT TAX	**$169.4**
Pass-through price increase as percentage of total sales	1.8%
Elasticity rate	0.65
Net impact of price increase on sales	-1.2%
Net decline in sales caused by eliminating subsidy	$2.0

The second support change affects the gross figure of $115.1 billion in food and nutrition assistance that the USDA provides to low-income Americans. My recommendation is to exclude animal foods from nutrition assistance programs. Applying the 32 percent multiplier introduced in chapter 5 (consumer spending on animal foods as a portion of total retail food spending) yields a total of $36.8 billion related to animal foods to be shifted to non-animal foods.[6] In a Meat Tax economy, because consumption of animal foods is lower by 32.5 percent, excluding animal foods from nutrition support represents a net decline in animal food sales of 67.5 percent of $36.8 billion, or $24.8 billion. This figure represents 14.6 percent of the total retail sales under the Meat Tax of $169.4 billion.

TABLE C3 Effects of Excluding Animal Foods from Nutrition Support (dollar amounts in billions)

TOTAL SUPPORT	$115.1
Portion related to animal foods	32%
Total support related to animal foods	$36.8
Percent of total consumed under Meat Tax	67.5%
Net consumption shift to non-animal foods	$24.8

Drop in Consumption

This proposal would cause annual US consumption of animal foods to fall by an estimated 44.1 percent, from $251 billion to $140 billion. It's easiest to explain this drop as occurring in multiple steps. First, with demand elasticity at 0.65, the 50 percent tax and resulting price increase causes consumption to drop by 32.5 percent (0.5 x 0.65)— yielding an initial sales decline from $251 billion to $169 billion. Next, as we saw above, excluding animal foods from nutrition support causes these sales to decline by a further 14.6 percent, eliminating check-offs causes a further sales decline of 1.8 percent,[7] and eliminating the research and marketing subsidy causes sales to further decline by 1.2 percent. The net result is that retail sales fall to about $140.3 billion, 44.1 percent below their current level. Note that there is no particular importance in the order of these steps; the net result of $140.3 billion is the same regardless of how the steps are sequenced.

TABLE C4 Effects of Meat Tax and Support Changes on Quantity Demanded (dollar amounts in billions)

ANNUAL RETAIL SALES OF ANIMAL FOODS	$251.0
Step 1: Sales decline caused by Meat Tax	32.5%
Retail sales under Meat Tax	$169.4
Step 2: Sales decline from excluding animal foods from nutrition support	14.6%
Retail sales after excluding animal foods from nutrition support	$144.6
Step 3: Sales decline from eliminating checkoffs	1.8%
Retail sales after eliminating checkoffs	$142.0
Step 4: Sales decline from eliminating research and marketing subsidy	1.2%
Net retail sales after Meat Tax and support changes	$140.3

New Tax Burden

The Meat Tax will impose a new tax burden on Americans. We can calculate this burden by determining consumption levels under the Meat Tax, and then determining the tax paid by consumers at those consumption levels. The US Bureau of Labor Statistics estimates that on average, American family units spend between $1,468 and $2,651 yearly on meat, eggs, dairy, and fish.[8] Table C5 shows the Meat Tax's estimated burden on Americans by family unit type, after accounting for projected declines in consumption.

TABLE C5 Meat Tax Burden on American Family Units

	MARRIED COUPLE HOUSEHOLD	SINGLE HEAD OF HOUSEHOLD	SINGLE
Annual spending on animal foods	$2,651	$1,745	$1,468
Percent demand decline with policy changes	44.1%	44.1%	44.1%
Net animal food spending under policy changes	$1,482	$975	$821
Meat Tax rate	50%	50%	50%
Annual Meat Tax payable	$741	$488	$410

Tax Credits

The proposal includes providing tax credits to offset the cost of the Meat Tax to individual taxpayers. Rounding each family unit's Meat Tax–related burden up to the nearest multiple of $10 to determine the appropriate credit amount, the proposed credits and projected tax relief are as shown in table C6.

TABLE C6 Tax Credits and Burdens by Family Unit

	MARRIED COUPLE HOUSEHOLD	SINGLE HEAD OF HOUSEHOLD	SINGLE
Tax credit	$750	$490	$410
Tax burden	741	488	410
Difference	$9	$2	$0

Using the Internal Revenue Service's figures for taxpayers by family unit type, the proposed tax credits would cost an estimated $78.4 billion, as shown in table C7.

TABLE C7 Cost of Tax Credits[9]

	TAX CREDIT	TAXPAYER UNITS (IN MILLIONS)	TOTAL (IN BILLIONS)
Married couple household	$750	56.1	$42.1
Single head of household	490	21.5	10.5
Single taxpayer	410	62.8	25.8
Grand total			$78.4

Lower Government Payments Under Medicare and Medicaid Programs

Another effect of the proposed changes is to lower state and federal expenditures on Medicare and Medicaid programs. Of all US health care expenditures, Medicare pays an average of 20 percent and Medicaid an average of 15 percent.[10] Accordingly, reductions in direct US health care costs associated with animal food consumption will lead to lower government payments under these programs, as shown in table C8. The federal government pays all Medicare costs and roughly two-thirds of Medicaid costs, yielding total annual federal savings of $22.4 billion. The total annual state savings are $3.6 billion. (Note that because these savings relate to gains in wellness that will be achieved only as individual eating habits lead to fewer health care problems, it may take several years of improvement in diet and health before these gains are fully realized each year.)

TABLE C8 Reduced Government Payments to Medicare and Medicaid Programs (dollar amounts in billions)

	HEART DISEASE	DIABETES	CANCER	TOTAL
Direct health care costs	$291.0	$128.4	$114.9	$534.3
Percent related to animal foods	30.0%	33.3%	33.3%	
Direct costs related to animal foods	$87.3	$42.8	$38.3	$168.4
Consumption decline under Meat Tax	44.1%	44.1%	44.1%	
Direct costs saved related to animal foods	$38.5	$18.9	$16.9	$74.3
Percent paid by Medicare	20.0%	20.0%	20.0%	
Direct costs saved by Medicare	$7.7	$3.8	$3.4	$14.9
Percent paid by Medicaid	15.0%	15.0%	15.0%	
Direct costs saved by Medicaid	$5.8	$2.8	$2.5	$11.1
Percent of Medicaid paid, federal	67.7%	67.7%	67.7%	
Medicaid costs saved, federal	$3.9	$1.9	$1.7	$7.5
Percent of Medicaid paid, state	32.3%	32.3%	32.3%	
Medicaid costs saved, state	$1.9	$0.9	$0.8	$3.6
Total federal savings				$22.4
Total state savings				$3.6
Total state and federal savings				$26.0

(Note that the *direct* health care costs itemized above are significantly lower than the *total* health care costs attributable to these three diseases. That's because the total costs include *indirect* costs such as lost wages, which are excluded from the above calculations as they do not affect Medicare or Medicaid.)

Change in US Treasury's Cash Flow

Using data from tables C2, C4, C7, and C8, we can estimate the annual impact these proposed changes will have on cash in the US Treasury. With adjusted retail sales of animal foods at $140.3 billion and a tax rate of 50 percent, the Meat Tax will generate tax revenue of about $70.2 billion. Combining this with the other estimated figures yields a total annual federal cash surplus of about $32.4 billion, as shown in table C9.

TABLE C9 Changes in US Treasury's Cash Flow (in billions)

MEAT TAX REVENUE	$70.2
Medicare and Medicaid savings	22.4
Savings from research and marketing subsidy	18.2
Gross additional cash	110.8
Less tax credit cost	-78.4
Net cash surplus	$32.4

Declines in Externalized Costs

Finally, the Meat Tax and other proposed changes would greatly reduce the externalized costs of meat and dairy production. As shown in table C10, the total annual reduction in externalized costs is estimated at roughly $184 billion. With the exception of subsidies, all externalized costs would be reduced by 44.1 percent, which is the factor by which the policy changes are estimated to reduce consumption and hence production. Subsidies are reduced by the amount of the research and marketing subsidy proposed to be eliminated.

TABLE C10 Annual Reduction in Externalized Costs of Animal Foods Resulting from Meat Tax and Other Changes (dollar amounts in billions)

	PRE-CHANGE LEVEL	REDUCTION FACTOR	POST-CHANGE LEVEL	NET REDUCTION
Health care	$314.0	-44.1%	$175.5	$138.5
Subsidies	38.4	-$18.2	20.2	18.2
Environment	37.2	-44.1%	20.8	16.4
Cruelty	20.7	-44.1%	11.6	9.1
Fishing	4.5	-44.1%	2.6	1.9
Total	$414.8		$230.7	$184.1

Factory Farming Practices

There is a common belief, at least among animal food producers, that technological progress in farming helps the animals whose lives it affects. Efficiency is good for economic markets, and as Dickens wrote, "A good thing can't be cruel."[1] Like most business operators, animal farmers welcome almost any innovation that improves efficiency and boosts profits. Just as it has for the car industry, repeated tinkering to improve processes and increase outputs has yielded significant productivity gains over the past century for animal agribusiness. As noted, the last century saw a tripling in per-cow dairy production and a doubling in per-hen egg production. These efficiency gains have been driven by advances in areas like feeding, handling, selective breeding, and of course, confinement methods. Yet while humans associate innovation and efficiency with progress and improvement, the animals, if they could talk, would almost certainly disagree. This appendix explores the production methods which prevail in factory farms across the United States and are routinely used in raising hogs, dairy cows, veal calves, broiler chickens, and laying hens.

It's a Hog's Life

In his wonderful nonfiction book *The Pig Who Sang to the Moon*, Jeffrey Masson reports the story of Lulu—a two-hundred-pound pig living at an animal sanctuary:

> Joanne Altsmann was in her kitchen one afternoon, feeling unwell, when Lulu charged out of a doggie door made for a 20-pound dog, scraping her sides raw to the point of drawing blood. Running into the street, Lulu proceeded to draw attention by lying down in the middle of the road until a car

stopped. Then she led the driver to her owner's house, where Altsmann had suffered a heart attack. Altsmann was rushed to the hospital, and the ASPCA awarded Lulu a gold medal for her heroism. Altsmann knows in her bones that Lulu's sixth sense saved her life.[2]

Pigs, Masson says, are sensitive, loyal, and intelligent. They're capable of forming complex social relationships, and they wag their tails like dogs when they're happy.

But in factory farms, where virtually all pigs in the United States are raised, hyper-confinement means that these and other animals lack the space or outdoor access to engage in instinctive behaviors. How do the animals like living in these conditions? Matthew Scully, former speech writer for George W. Bush and author of the book *Dominion: The Power of Man, the Suffering of Animals, and the Call to Mercy,* sought a firsthand answer to that question. Through Scully, we learn from North Carolina pig farmer F. J. "Sonny" Faison how pigs feel about spending their lives in what Faison calls "state-of-the-art confinement facilities." According to Faison, the animals "love it. . . . They don't mind at all. . . . The conditions . . . are much more humane than when they were out in the field."[3]

Another North Carolina pig farmer, Jerry Godwin, also extolled his plant's modern methods to Scully: "If you want to look at an animal in one of our systems, at the way it is housed, you look at that and say, 'Oh my gosh, that's terrible.' Well, the fact is that to that animal it may not be so bad. That animal seems to live longer, to prosper, to do well. Its comfort is there."[4]

But Scully's impressions while touring a hog farm didn't support pig farmers' claims that innovation in porcine agriculture benefits the animals. At a supposedly state-of-the-art hog factory in North Carolina, Scully saw "sores, tumors, ulcers, pus pockets, lesions, cysts, bruises, torn ears, swollen legs everywhere. Roaring, groaning, tail biting, fighting, and other 'vices,' as they're called in the industry. Frenzied chewing on bars and chains, stereotypical 'vacuum' chewing on nothing at all, stereotypical rooting and nest building with imaginary straw."[5] Scully was invited to tour that particular facility by Sonny Faison, the pig farmer who said his animals "love" their living conditions. In truth, as we see repeatedly in this example and others in this appendix, innovation in animal farming often means a backward step in the animals' quality of life. Notwithstanding claims to the contrary by factory farm operators, the evidence shows that when the focus turns to raising an animal faster, on cheaper feed, or in less space, the animal invariably loses.

Dairy's Dark Side

As a child visiting his uncle's farm in Wisconsin, physician Michael Klaper saw a dairy cow separated from her newborn calf. The incident left a lasting impression. Years later, he wrote:

> The mother was allowed to nurse her calf but for a single night. On the second day after birth, my uncle took the calf from the mother and placed him in the veal pen in the barn—only ten yards away, in plain view of the mother. The mother cow could see her infant, smell him, hear him, but could not touch him, comfort him, or nurse him. The heartrending bellows that she poured forth—minute after minute, hour after hour, for five long days—were excruciating to listen to. They are the most poignant and painful auditory memories I carry in my brain.[6]

There's a popular belief, long outdated, that dairy cows lead blissful lives. But as this and other examples show, life on a dairy farm is anything but easy.

Centuries or even decades ago, life might have been different for a dairy cow. But with the typical dairy farm's footprint changing from pastoral to industrial, production methods have changed too. Today, while dairy cows would otherwise live past twenty, they're generally killed for beef before the age of four.[7] Further, contrary to the conventional wisdom, cows don't routinely make milk and they don't *need* to be milked—like humans, they lactate only after giving birth. Unlike most humans, however, they're forcibly inseminated a number of times during their lives.[8]

As Klaper saw as a child, calves must be separated from their mothers within hours of birth; otherwise, the maternal bond grows too strong and makes separation especially difficult. Of course, even an immediate separation is painful for a mother whose mammary glands are designed to feed her own young and whose most basic instinct is to do so. And what happens next, following separation, depends on the calf's sex.

Veal Calves

Most males born in the dairy industry are destined for veal crates—tiny stalls banned in the European Union but permitted in most of the United States. As John Robbins, author of *Diet for a New America,* observed, "The veal calf would actually have more space if, instead of chaining him in such a stall, you stuffed him into the trunk of a subcompact car and kept him there for his entire life."[9] For the connoisseur, veal's appeal lies in its softness and paleness. Thus, calves are tethered to prevent any but the slightest movement—this immobility keeps the infants' flesh tender by preventing muscle development. To keep them anemic and maintain their flesh's characteristic pink color, newborns are denied their mothers' milk and instead fed formula without iron. The young males are typically slaughtered at four months.

The inhumane treatment of veal calves is no mystery to most American consumers, who, since learning about veal in the mid-1970s, have responded by dramatically reducing their consumption of the anemic flesh. From 1975 to 1998, annual US per capita veal consumption fell 77 percent from almost 4 pounds to less than 1.[10] Yet in counterpoise to the veal industry's decline, dairy consumption provides this dying industry with endless rebirth. Almost one in two calves born to dairy cows every day lands in a confinement crate, destined to be marketed to veal eaters in the United States or abroad. This seems a particularly bizarre irony for the millions of US consumers who would not dream of eating veal but who, by consuming dairy, power an industry that many believe should have died long ago.

Battery Cows

Female calves, on the other hand, are destined for a life of milk production. Dairy's innovative answer to the battery cage is zero grazing, a system of intensive confinement that keeps cows tethered in stalls—usually of steel and concrete—for most of their lives. Unlike conventional dairy farming, which relies on pasture, zero grazing requires little land and is thus scalable in ways that pasture grazing is not. The rise of zero grazing over the past several decades has led to a heavy drop in the number of dairy farms and a sharp increase in the cow population at those that remain. Between 1970 and 2006, the number of US farms with dairy cows fell from 648,000 to 75,000. With this consolidation in the industry, the majority of US milk is now produced on farms with five hundred or more cows—nearly all of which are zero grazing.[11]

Yet cows, just like people and other animals, enjoy the wind in their hair and the grass under their feet. Research shows that given a choice, cows spend the majority of their time outdoors and choose to come inside only to escape high temperatures.[12] As one dairy worker observed, "The thing you notice with zero grazing is how depressed and uptight the cows are. The eyes are dull."[13] Because of the parallels to intensely confined laying hens, some call these living milk machines "battery cows."

Cows might look dull, but don't be fooled. In fact, research shows cows are smarter than we thought. It also finds they're capable of feeling deep emotions and forming complex social relationships. In one study, cows were challenged to open a door to find food while their brain waves were measured. When they solved the problem, they felt a thrill, according to Donald Broom, the Cambridge University professor who led the study. "The brainwaves showed their excitement; their heartbeat went up and some even jumped into the air. We called it their Eureka moment," Broom said.[14] In another study, researchers at Bristol University found that cows typically form friendships with two to four other animals and spend most of the time with their friends.[15] Like many people, they may dislike others of their species and bear grudges for years.

The Chicken and the Egg

"If you grew as fast as a chicken," according to the University of Arkansas Division of Agriculture, "you'd weigh 349 pounds at age two."[16] Broiler chickens—so-called because they yield meat, not eggs—are bred to get as big as possible as fast as possible. They now grow twice as fast and get more than twice as big as they once did, prompting the awful pun "double broiler." The rapid growth and distorted body size of broiler chickens means their legs and organs can't keep pace with the rest of their body, often leading to disease and deformity. According to one published study, "Broilers now grow so rapidly that the heart and lungs are not developed well enough to support the remainder of the body, resulting in congestive heart failure."[17]

They can't walk so well either. Ninety percent of broiler chickens have abnormal gaits caused by genetic bone deformities.[18] The pain of these deformities leads chickens to dose themselves with pain medicine (if available), by consistently choosing feed containing anti-inflammatory drugs over regular feed.[19] "Broilers," wrote Professor John Webster of the University of Bristol School of Veterinary Science, "are the only livestock that are in chronic pain for the last 20 percent of their lives. They don't move around, not because they are overstocked, but because it hurts their joints so much."[20] Six-week-old broilers have such a hard time supporting their abnormally heavy bodies that they spend up to 86 percent of the time lying down.[21] And their constant contact with ammonia-laden litter leads to burns, breast blisters, and foot pad dermatitis.[22]

Yet for all the difficulties in a broiler chicken's life, conditions are no better for their hardworking cousins—laying hens. In fact, because of the way hens and their offspring are treated, eggs—a dietary staple for even many a vegetarian—are surprisingly inhumane. As factory farm critic Erik Marcus writes, "A bite of egg involves more animal suffering than a bite of hamburger or bacon."[23]

Life in the Industrial Henhouse

For chickens in the laying industry, life starts inauspiciously. Male chicks are useless because they cannot lay eggs and, unlike genetically engineered broiler chickens, are not bred for the rapid growth that makes it profitable to produce chicken meat. With no laws or humane standards mandating how unwanted chicks must be handled, farm operators are left to discard the day-old birds in whatever manner is most cost-effective. This could mean shredding them alive in a meat grinder or wood chipper, dumping them in a garbage can to starve to death, or stuffing them in a garbage bag to suffocate. Egg producers kill 270 million unwanted male chicks each year, enough tiny dead birds to circle the contiguous United States.[24] Nevertheless, if they knew what was in store for their sisters, these baby roosters might be grateful for their early deaths.

For chickens unlucky enough to be born female and destined for a laying career, life starts with a painful surgical procedure known clinically as "partial beak amputation." Euphemistically called "beak trimming" by those in the industry, this procedure involves cutting off about one-third of an unanesthetized chick's beak and leaving the sensitive nerve endings exposed for the remainder of her life. As one group of researchers explains:

> The avian beak is a complex sensory organ which not only serves to grasp and manipulate food particles prior to ingestion, but is also used to manipulate non-food articles in nesting behavior and exploration, drinking, preening, and as a weapon in defensive and aggressive encounters. To enable the animal to perform this wide range of activities, the beak of the chicken has an extensive nerve supply with numerous [nerve endings sensitive to pressure, heat, and pain]. . . . Beak amputation results in extensive neuromas [tumors] being formed in the healed stump of the beak which give rise to abnormal spontaneous neural activity in the trigeminal nerve. . . . Therefore, in terms of the peripheral neural activity, partial beak amputation is likely to be a painful procedure leading not only to phantom and stump pain, but also to other characteristics . . . such as hyperalgesia [extreme sensitivity to pain].[25]

There is no human analog to debeaking, though it's not much of a stretch to compare it to having your lips chopped off and cauterized, with the exposed flesh and nerve endings left to react painfully to whatever you eat, drink, or touch with your lips for the rest of your life. Researchers who monitored hens' pecking, drinking, beak-wiping, and head-shaking activities after being debeaked observed significant changes in these activities that persisted long after the hens' beaks appeared to be healed. The researchers blamed these persistent behavioral changes on the hens' increased sensitivity to pain and concluded, "The modifications in the pecking and drinking behavior of birds following partial beak amputation [conforms with other reports] that partial beak amputation results in long-term increases in dozing and general inactivity, behaviors associated with long-term chronic pain and depression."[26]

Like most other painful mutilations to which farmed animals are subjected, debeaking has only the farm operator's bottom line in mind. Birds in close confinement would normally peck one another to death because of stress, but once debeaked, they find it too painful to do so. It's like cutting off a human inmate's knuckles to stop him from punching others.

As an innovation, debeaking started in 1940. That's when a San Diego poultry farmer discovered he could stop his chickens from pecking each other by burning off their upper beaks with a blow-torch.[27] Further "advances" led to the use of a searing blade, the current technology, to debeak birds instead of a blowtorch. But for laying hens, debeaking is just the beginning.

Innovation in Egg Production

Battery means either "a number of similar things occurring together" or "offensive physical contact or bodily harm."[28] In the case of the battery cage used to house laying hens, the word takes on an apt double meaning. Fold down three inches at the top of a regular sheet of paper to leave a smaller sheet measuring 8.5″ by 8″, or 68 square inches. Most laying hens spend their entire two-year lifespan in a space that size or smaller—the bottom of an industry-standard cage is 67 square inches. To get an idea of the conditions in a typical battery cage, imagine ten birds living in the drawer of a filing cabinet. While many people would consider this cramped, the National Chicken Council assures us that confining laying hens in such cages "is the most effective way to keep [them] comfortable and in good health."[29]

First patented in 1909, battery cages were not originally intended to house birds for their entire lives but only to rear chicks during the "dangerous early stages."[30] However, when chicken farmers discovered the cages could be used to house chickens indefinitely, their use quickly proliferated. One early enthusiast was Milton Arndt, a sales manager in the brooding division of Kerr Chickeries in New Jersey. Arndt conducted experiments to determine whether it was more profitable to house laying hens individually or in batteries. In 1931, he wrote, "Birds confined in the batteries outlaid considerably the same size flock in the regular houses. The birds consume less feed than those on the floor and this coupled with the increased production made them more profitable than the same number of pullets in the laying house."[31]

Of course, the droppings have to go somewhere, which is why the cages have wire mesh floors. This works well for the farm operators whose clean-up duties are simplified, but not so well for the birds whose entire lives are spent standing on the unnatural wire surface. When animal advocates speak to school groups about the conditions in egg factories, they sometimes bring a wire mesh surface for students to stand on, barefooted, to appreciate what the hens experience. Standing on these surfaces for their whole existence, birds often develop painful joint conditions, brittle bones, and crippling deformities. By the end of their lives, 30 percent of laying hens are likely to have broken bones.[32]

Forced Molting

Hens in the wild molt, or replace their feathers, annually at the end of their laying season (generally in the fall). Wild hens typically lay about twenty eggs per year. By contrast, hens in US factory farms laid an average of 269 eggs in 2010, slightly more than needed to satisfy a typical American's annual consumption of 246 eggs.[33] This thirteenfold productivity increase in egg factories is driven by innovations such as forced molting, the practice of starving hens for up to two weeks to increase productivity. Hens forced to molt typically lose one-third of their weight but, because of physiological changes relating to fertility, become better egg layers when it's over.

What's it like to be starved for two weeks? For laying hens, according to one researcher, it causes "extreme distress" evidenced by numerous physiological and psychological changes, including "increased aggression and . . . pacing."[34] According to United Poultry Concerns, whose exposure of forced molting helped bring the issue to the forefront, starving hens become so desperate for food, they eat one another's feathers.[35] Captive hens' normal mortality rate, about 15 percent per year, doubles during forced molting. The American Egg Board recommends that hens be forced to molt twice in their lives: at fourteen months and twenty-two months of age.[36] Because a third episode of forced molting typically wouldn't yield sufficient productivity increases to be worthwhile, the "spent" hens are slaughtered instead at about age two. Otherwise, hens would have a lifespan of ten years or longer.[37]

The Uncontrollable Urge to Act Like a Chicken

In the wild, hens roost and build nests in safe places. They spread and flap their wings. They take frequent dust baths to remove insects and debris from their feathers. They scratch for food, because often the best seeds or insects are found under leaves or other covering. Intelligent, social animals, they bond with their young and with other members of their flock.[38] Chickens have evolved these behaviors over millions of years, and they are driven to engage in them whether they live in the wild or in confinement.

Categorically denied the natural behaviors coded into their DNA, battery hens nevertheless go through the unsatisfying motions their bodies crave. These thwarted impulses have a particular pathos, a futility and frustration of purpose that observers say is hard to watch. "The worst torture to which a battery hen is exposed is the inability to retire somewhere for the laying act," wrote the late Nobel Prize–winning zoologist Konrad Lorenz. "[I]t is truly heart-rending to watch how a chicken tries again and again to crawl beneath her fellow-cage mates to search there in vain for cover."[39] Another researcher writes of the animals' frustrated desire to dust-bathe:

> Chickens in battery cages which have wire floors and no loose substrate for the birds to scratch and dust bathe in can often be seen to go through all the motions of having a dust bath. They squat down, raise their feathers, and rub themselves against the floor and flick imaginary dust from their backs. They behave as though real dust were being moved through their feathers, but there is nothing really there. If such dust-deprived birds are eventually given access to something in which they can have a real dust bath, like wood shavings or peat, they go in for a complete orgy of dust bathing. They do it over and over again, apparently making up for lost time.[40]

A free hen under attack uses a unique call to summon help from her rooster. Maybe because she doesn't want to dilute the call's efficacy by crying wolf, she never uses it for any other purpose. Caged battery hens make the same last-resort call for help.[41] This means the cacophony in a henhouse is much more than purposeless noise: it's actually the sound of thousands of distressed hens repeating a rescue call over and over.[42]

Cage-Not-So-Free

Americans' hearts are increasingly in the right place, although sometimes we still lack facts. The American Egg Board estimates that at least 5 percent of the US eggs consumed in 2010 were cage-free, a number that grows as consumers are egged on to buy the output of "humanely" raised hens.[43] We're told cage-free eggs are a compassionate alternative to battery eggs because, as one "progressive" farm boasts, the hens have "plenty of room to do the things that hens love most: scratch, flap their wings, perch, nest and roost in a carefully managed, safe, low stress environment."[44] However, while this may be the case at a handful of small, alternative farms, it is not the case at the vast majority of cage-free facilities.

In fact, most cage-free hens are raised in industrial environments identical in almost all respects to battery cage facilities. Thus, like battery hens, cage-free hens are typically raised in dark steel-and-concrete warehouses reeking of ammonia and other fumes, where they are denied sunlight, dirt in which to bathe or scratch, and straw or other materials in which to nest. They're subjected to partial beak amputation and forced molting. Their brothers are killed at birth and discarded en masse in any manner that's cheap and easy.

The term *free-range* refers to eggs from hens with access to the outdoors. However, few birds take advantage of the ability to go outside. Instead, free-range chickens follow learned behaviors and stay inside, generally ignoring or avoiding the outdoors. As noted, Michael Pollan observed that during his visits to free-range chicken farms, he never actually saw a bird go outside.[45]

Jewel Johnson runs the Peaceful Prairie Sanctuary, a haven for rescued farm animals in Deer Trail, Colorado. She writes of her visit to a huge henhouse at a well-known organic, free-range egg farm:

> There was a strong stench . . . like a chicken coop times 10,000.
> . . . [The hens'] beaks were chopped off at the end. Their necks were featherless. Their combs were pale skin color untouched by the sun. . . . These birds only had the grate they were standing on and the metal walls surrounding them until they died.
> . . . There was no straw, and there was no wood to perch on. There was nothing natural in that building other than death and suffering. There were no windows to see a world other than this. The only roost was a metal one designed to collect eggs and take them away from the birds. There was nothing to build a nest with unless the birds used their feces and lost feathers as building material.[46]

That's "humane" egg farming in an eggshell. The industry takes a loosely defined standard like "free-range" or "cage-free," determines the minimum it must do to meet that standard, and proceeds accordingly. This kind of cage-free egg production might be incrementally better than a battery cage system, but is it really humane?

To ask the question another way, is it humane for cage-free hens to be debeaked, denied basic instincts, assaulted by caustic fumes, subjected to starvation bouts, and crowded by the tens of thousands into dark warehouses?

The Secret Lives of Chickens

Some people may be inclined to dismiss details like these on the grounds that chickens are bird-brained and not smart enough to care. But remove any random chicken from its industrial environment, as rescuers sometimes do, and you'd soon discover an intelligent, social companion with a unique and friendly personality. Jeffrey Masson, who has written about his experience as friend and guardian to two rescued chickens, said the animals are "funny, curious, affectionate, stubborn, ingenious companions."[47]

A recent study suggests we can add empathetic to this list of traits. Researchers at the University of Bristol placed hens and their chicks in separate enclosures where the hens could see, smell, and hear their young. They subjected the chicks to short puffs of compressed air, which caused mild discomfort but no real pain. The chicks reacted aversively but did not make a distress call. Nevertheless, seeing their chicks apparently suffering, hens experienced an empathetic stress response that elevated their heart rates and body temperatures and made them vocalize.[48]

Studies find that chickens are as smart as mammals, including some primates.[49] In an interview with Chris Evans, a chicken researcher at Macquarie University in Australia, a *New York Times* reporter noted, "The chicken [has an] intriguing ability to understand that an object, when taken away and hidden, nevertheless continues to exist. This is beyond the capacity of small children."[50] And Colorado State University Professor Bernard Rollin observes, "Contrary to what one may hear from the industry, chickens are not mindless, simple automata but are complex behaviorally, do quite well in learning, show a rich social organization, and have a diverse repertoire of calls. Anyone who has kept barnyard chickens also recognizes their significant differences in personality."[51]

Karen Davis runs a poultry sanctuary in Machipongo, Virginia, founded the advocacy group United Poultry Concerns, and literally wrote the book on chicken welfare (*Prisoned Chickens, Poisoned Eggs: An Inside Look at the Modern Poultry Industry*). Writing of her experiences living with dozens of chickens rescued from industrial farming, Davis observes:

> Chickens represented by the poultry industry as incapable of friendship with humans have rested in my lap with their eyes closed as peacefully as sleeping babies, and . . . they quickly learn their names. A little white hen from the egg industry named Karla became so friendly, all I had to do was call out "Karla!" and she would break through the other hens and head straight toward me, knowing she'd be scooped off the ground and kissed on her sweet face and over her closed eyes. And I can still see Vicky, our large white hen from a "broiler breeder" operation, whose right eye had been knocked out, peeking around the corner of her house each time I shouted, "Vicky, what are you doing in there?" And there was Henry, likewise from a broiler breeder operation, who came to our sanctuary dirty and angry after falling out of a truck on the way to a slaughter plant. Lavished with my attention, Henry, who at first couldn't bear to be touched, became as pliant and lovable as a big shaggy dog. I couldn't resist wrestling him to the ground with bearish hugs, and his joy at being placed in a garden where he could eat all the tomatoes he wanted was expressed in groans of ecstasy.[52]

Industrial animal farmers don't set out to ignore animals' needs. Like other factory operators, they just want to keep costs low. But unlike a car maker or a toy company, the animal food industry's production units are living beings whose quality of life depends almost entirely on the amount spent on their welfare. There will always be conflict between making money and raising animals humanely. In any given year, one in seven laying hens dies of easy-to-treat causes like starvation, dehydration, or a prolapsed uterus. These deaths are just part of the cost of doing business, but the individual suffering that precedes each death doesn't show up in the financial statements. Why bother to provide costly veterinary care, when a laying hen can be replaced for $3? Why house birds comfortably in two sheds when they can be maintained more cheaply—albeit stressfully tightly—in one? Why settle for reduced egg output when hens can produce another dozen or two if they're starved to extreme distress?

Enriched Cages

Increasingly, voters and lawmakers around the world are questioning the egg industry's confinement practices. The European Union banned battery cages in 2012, requiring that laying hens live either cage-free or in enriched cages, which provide more space per bird as well as enrichment devices such as perches, nest boxes, and scratching areas. California's Proposition 2 requires that from 2015, laying hens must be housed in cages large enough to let them fully extend their wings in all directions without touching a cage wall or another hen.[53] And in the wake of an agreement to support enriched cages between the United Egg Producers and the Humane Society of the United States, it seems that Congress may soon adopt legislation requiring enriched cages for hens in the United States.[54]

However, some critics of the egg industry argue that enriched cages are not enough, and the only solution is to eliminate cages altogether. One commentator is veterinarian and University of California Professor Emeritus Nedim C. Buyukmihci, who writes:

> The increase in cage size dictated by [proposed enriched cage legislation], unfortunately, will have no meaningful positive impact. . . . Hens will still not be able to get proper exercise, they still will be too crowded to even properly stretch their wings, perches will be at an ineffectual height, and nest boxes will not be conducive to the needs for laying eggs.[55]

Such debate is not unusual among those concerned for farm animals' welfare. Many believe, with good reason, that so-called humane farming measures do little to protect animals, and they'd rather see the abolition, not the amelioration, of the cruel practices found in factory farms. As we've seen, when implemented by industrial methods, even farming practices labeled *organic, cage-free,* and *free-range* are routinely little better for the animals than the more blatantly inhumane alternatives. For that reason, while I believe that eating less animal foods—or giving them up altogether—is a good way for an individual to address the problems described in this appendix, I don't advocate merely switching to purportedly humane animal products as a solution.

ENDNOTES

Author's Note

1   Shunryu Suzuki, *Zen Mind, Beginner's Mind: Informal Talks on Zen Meditation and Practice* (New York: Weatherhill, 1989).

Introduction

1.   Allan Schinkel, "Pork Production Costs: Farrow to Finish Production," *Animal Sciences* 443: Swine Management (2000), accessed December 1, 2011, *http://www.ansc.purdue.edu*.

2.   The loss per animal of $20 to $90 is for larger and more efficient producers—that is, those raising 100 or more head of cattle. Smaller producers' losses are even higher, ranging from $184 to $305 per animal. Sara D. Short, "Characteristics and Production Costs of U.S. Cow-Calf Operations," *USDA Statistical Bulletin* 17, no. 947–3 (2001), accessed December 1, 2011, *http://www.ers.usda.gov*.

3.   *See* chapter 5.

4.   Christopher Chantrill, "Government Spending Details," US Government Spending (2012), accessed July 10, 2012, *http://www.usgovernmentspending.com*.

5   The 2013 farm bill (not yet passed as of this writing) seeks to discontinue such direct payments. Michael Grunwald, "Why Our Farm Policy is Failing," *Time Magazine* (November 2, 2007).

6.   Boris Worm et al., "Impacts of Biodiversity Loss on Ocean Ecosystem Services," *Science* 314, no. 5800 (2006): 787–90.

7.   Jeff Herman, "Saving U.S. Dietary Advice from Conflicts of Interest," *Food & Drug Law Journal* 65 (2010): 285–326.

8.   *See* chapter 4.

9.   *See* chapter 4.

10. Robert Goodland and Jeff Anhang, "Livestock and Climate Change: What if the Key Actors in Climate Change Are . . . Cows, Pigs and Chickens?" *World Watch* (November/December 2009): 10–19, accessed October 25, 2011, *http://www.worldwatch.org*.

11. Herbert T. Buxton and Dana W. Kolpin, "Fact Sheet FS-027-02, Pharmaceuticals, Hormones, and Other Organic Wastewater Contaminants in U.S. Streams," US Geological Survey (2002), accessed October 24, 2011, *http://toxics.usgs.gov*.

12. *See* chapter 7.

13. Will Tuttle, *The World Peace Diet: Eating for Spiritual Health and Social Harmony* (New York: Lantern Books, 2005), xv.

14. Excluding the tiny state of Luxembourg, population 500,000, which apparently eats more meat per capita than we do but is too small to be statistically significant.

15. US Centers for Disease Control and Prevention, "U.S. Obesity Trends," accessed December 27, 2011, *http://www.cdc.gov*; World Cancer Research Fund International, "Data Comparing More and Less Developed Countries," accessed December 27, 2011, *http://www.wcrf.org*; American Cancer Society, "Cancer Facts and Figures 2011," accessed December 27, 2011, *http://www.cancer.org*; National Cancer Institute, "Surveillance Epidemiology and End Results," accessed December 27, 2011, *http://seer.cancer.gov*; World Diabetes Foundation, "Diabetes Facts," accessed December 27, 2011, *http://www.worlddiabetesfoundation.org*; American Diabetes Association, "Diabetes Statistics," accessed December 27, 2011, *http://www.diabetes.org*.

16. World Health Organization, "The World Health Report" (2000), accessed February 29, 2012, *http://www.who.int*.

17. *See* Appendix B.

18. *See* chapter 5.

19. *See* chapter 6.

20. Joe L. Outlaw et al., "Structure of the U.S. Dairy Farm Sector," *Dairy Markets and Policy: Issues and Options* (March 1996), accessed September 19, 2012, *http://aede.osu.edu*; US Department of Agriculture, "Overview of the United States Dairy Industry" (2010), accessed September 19, 2012, *http://usda.mannlib.cornell.edu*.

21. Farm Forward, "Factory Farming," accessed October 25, 2012, *http://www.farmforward.com*.

22. USDA Economic Research Service, "USDA Long-term Projection" (2007), accessed November 10, 2011, *http://www.ers.usda.gov*.

23. US Department of Agriculture, "Red Meat, Poultry, and Fish (Boneless Weight): Per Capita Availability" (2012), accessed September 19, 2012, *http://www.ers.usda.gov*.

24. National Center for Health Statistics, "Prevalence of Overweight, Obesity and Extreme Obesity Among Adults: United States, Trends 1976–80 through 2005–2006," Health E-Stats (December 2008).

25. Ibid.

26. Stephen Ansolabehere, John de Figueiredo, and James M. Snyder Jr., "Why Is There So Little Money in U.S. Politics?" *Journal of Economic Perspectives* 17, no. 1 (2003): 105–130; Center for Responsive Politics, "Money Wins Presidency and 9 of 10 Congressional Races in Priciest U.S. Election Ever" (2008), accessed July 10, 2012, *http://www.opensecrets.org*.

27. US Senate Office of Public Records, "Lobbying Disclosure Act Databases," accessed May 5, 2012, *http://www.senate.gov*.

28. Melanie Joy, *Why We Love Dogs, Eat Pigs, and Wear Cows: An Introduction to Carnism* (San Francisco: Conari Press, 2010).

29. Henning Steinfeld, "The Livestock Revolution—A Global Veterinary Mission," *Veterinary Parasitology* 125, nos. 1–2 (2004): 1–4.

30. Marta G. Rivera-Ferre, "Supply vs. Demand of Agri-Industrial Meat and Fish Products: A Chicken and Egg Paradigm?" *International Journal of the Society of Agriculture & Food* 16, no. 2 (2009): 90–105.

Chapter 1

1. Kentucky Cattlemen's Association, "US Federal Income Tax Return," 2009, accessed April 25, 2012, *http://www.guidestar.org.*

2. David Shipman, "Industry Insight: Checkoff Programs Empower Business," *USDA Blog* (2011), accessed December 31, 2011, *http://blogs.usda.gov.*

3. USDA Agricultural Marketing Service, "Benefits of Research & Promotion Boards (Checkoffs)" (2011), accessed October 27, 2012, *http://www.ams.usda.gov.*

4. Ibid.

5. Ibid.

6. Dairy Management, Inc., "Dairy Checkoff Highlights" (2011), accessed January 3, 2012, *http://www.dairycheckoff.com.*

7. Dairy figure includes both "dairy products" and "fluid milk." Geoffrey S. Becker, "Federal Farm Promotion ('Check-Off') Programs," Congressional Research Service Report for Congress (2008), accessed November 5, 2011, *http://www.nationalaglawcenter.org.*

8. Ibid.

9. *Johanns v. Livestock Mktg. Ass'n* (2005) 544 U.S. 550.

10. Ibid., 560 61.

11. In 2001, the US Supreme Court refused to compel dissenting mushroom farmers to support the majority message of the mushroom checkoff program. The court held the mushroom checkoff violated the First Amendment because it merely imposed marketing requirements with little other regulation and hence was "not part of a comprehensive statutory agricultural marketing program." *United States v. United Foods, Inc.* (2001) 533 U.S. 405.

12. Chanjin Chung and Emilio Tostao, "Will the Voluntary Checkoff Program Be the Answer? An Analysis of Optimal Advertising and Free-Rider Problem in the U.S. Beef Industry," Southern Agricultural Economics Association (2004), accessed May 3, 2012, *http://ageconsearch.umn.edu.*

13. USDA Agricultural Marketing Service, "Benefits of Research & Promotion Boards (Checkoffs)" (2011), accessed January 26, 2012, *http://www.ams.usda.gov.*

14. Ibid., 5; Becker, "Federal Farm Promotion ('Check-Off') Programs."

15. Researchers use 0.77 as a typical multiplier to measure the effect on farm communities of an increase in jobs or income. Curtis Braschler et al.,

"Economic Base Multipliers and Community Growth," University of Missouri Extension (1993), accessed January 26, 2012, *http://extension.missouri.edu.*

16. The dairy category, for which no data are given, is assumed to have the same return on invested funds as fluid milk. USDA Agricultural Marketing Service, "Benefits of Research & Promotion Boards (Checkoffs)" (2011); Geoffrey S. Becker, "Federal Farm Promotion ('Check-Off') Programs.

17. Because the United States does not publish child-related cholesterol guidelines, the EFSA guidelines are used for this purpose. USDA Agricultural Research Service, "Nutrient Intakes from Food: Mean Amounts Consumed per Individual, One Day, 2005–2006" (2008), accessed January 26, 2012, *http://www.ars.usda.gov*; US Food and Drug Administration, "Calculate the Percent Daily Value for the Appropriate Nutrients," accessed January 26, 2012, *http://www.fda.gov*; European Food Safety Authority, "Scientific Opinion on Dietary Reference Values for Fats, Including Saturated Fatty Acids, Polyunsaturated Fatty Acids, Monounsaturated Fatty Acids, Trans Fatty Acids, and Cholesterol," *EFSA Journal* 8, no. 3 (2000): 30, accessed January 26, 2012, *http://www.efsa.europa.eu.*

18. American Heart Association, "Overweight in Children," accessed January 26, 2012, *http://www.heart.org.*

19. Dairy Management, "Dairy Checkoff Highlights."

20. US Department of Agriculture, "Benefits of Research & Promotion Boards."

21. Ibid.

22. Ibid.

23. National Dairy Council, "Research," *The Dairy Connection,* accessed September 20, 2011, at *http://www.nationaldairycouncil.org.*

24. According to the ASN website, sponsorship provides a corporation with "access to more than 12,000 scientists and practitioners." American Society for Nutrition, "ASN Sustaining Members," accessed September 20, 2011, at *http://www.nutrition.org*; American Dietetic Association, "2010 Annual Report," accessed December 14, 2011, *www.eatright.org.*

25. American Dietetic Association, "American Dietetic Association Welcomes National Dairy Council as an ADA Partner in the Association's New Corporate Relations Sponsorship Program," Press Release (March 7, 2007), accessed September 20, 2011, *http://www.eatright.org.*

26. Ibid.

27. Center for Science in the Public Interest, "Non-Profit Organizations Receiving Corporate Funding: American Heart Association," *Integrity in Science: A CSPI Project* (2006), accessed September 20, 2011, at *http://www.cspinet.org.*

28. Joel Lexchin et al., "Pharmaceutical Industry Sponsorship and Research Outcome and Quality: Systematic Review," *British Medical Journal* 326 (2003): 1167; Anastasia L. Misakian and Lisa A. Bero, "Publication Bias and Research

on Passive Smoking," *Journal of the American Medical Association* 280, no. 3 (1998): 303–4.

29. Lexchin et al., "Pharmaceutical Industry Sponsorship," abstract.

30. Patty W. Siri-Tarino et al., "Meta-Analysis of Prospective Cohort Studies Evaluating the Association of Saturated Fat with Cardiovascular Disease," *American Journal of Clinical Nutrition* 91, no. 3 (2010): 535–46.

31. *See, for example,* Rashmi Sinha et al., "Meat Intake and Mortality: A Prospective Study of Over Half a Million People," *Archives of Internal Medicine,* 169, no. 6 (2009): 562–71; Teresa T. Fung et al., "Prospective Study of Major Dietary Patterns and Stroke Risk in Women," *Stroke* 35 (2004): 2014–19; A. R. P. Walker, "Diet in the Prevention of Cancer: What Are the Chances of Avoidance?" *The Journal of the Royal Society for the Promotion of Health* 116, no. 6 (1996): 360–66; Romaina Iqbal, Sonia Anand, and Stephanie Ounpuu, "Dietary Patterns and the Risk of Acute Myocardial Infarction in 52 Countries: Results of the INTERHEART Study," *Circulation* 118, no. 19 (2008): 1929–37.

32. *See, for example,* American Meat Institute, "Myth: Americans Eat Too Much Meat and Its Saturated Fat Content Leads to Heart Disease," accessed October 4, 2011, *http://www.meatmythcrushers.com.*

33. *See* chapter 6, table 6.1.

34. Ibid.

35. T. Colin Campbell and Thomas M. Campbell, *The China Study* (Dallas: BenBella Books, 2004), Kindle edition.

36. Ibid.

37. Siri-Tarino et al, "Association of Saturated Fat with Cardiovascular Disease," 535, note 4.

38. Michael Wohlgenant et al., "Returns to Pork Producers from Marketing and Production Research," The Research Committee on Commodity Promotion (2008), accessed January 4, 2012, *http://commodity.dyson.cornell.edu.*

39. Commodity Promotion and Evaluation, 7 U.S.C. § 7401(b)(7).

Chapter 2

1. Chris Welch, "Inaccurate 'Swine' Flu Label Hurts Industry, Pork Producers Say," *CNN Health* (April 30, 2009), accessed April 19, 2012, *http://articles. cnn.com.*

2. Thomas H. Maugh II, "Swine Flu Danger Appears to Be Ebbing," *Los Angeles Times* (March 19, 2010).

3. Caitlin Taylor, "Obama Administration: Out with the 'Swine,' In with the 'H1N1 Virus,'" *ABC News Political Punch* (April 29, 2009), accessed September 15, 2011, *http://abcnews.go.com.*

4. Ibid.

5. Gavin J. D. Smith et al., "Origins and Evolutionary Genomics of the 2009 Swine-Origin H1N1 Influenza A Epidemic," *Nature* 459 (2009): 1122–25.

6. PolitiFact.com, "Don't Call It Pink Slime, Georgia Official Says," accessed April 20, 2012, *http://www.politifact.com.*

7. Ross Boettcher, "BPI Halts Production at Three Plants," *Omaha World Herald* (March 26, 2012).

8. 2009 US Federal Income Tax Return for National Cattlemen's Beef Association; 2010 US Federal Income Tax Returns for National Pork Council, American Meat Institute, National Meat Association, National Chicken Council, National Turkey Federation, US Poultry and Egg Association, United Egg Association, National Milk Producers Federation, and Western United Dairymen, accessed April 25, 2012, *http://www.guidestar.org.*

9. US Poultry & Egg Association, "Economic Data," accessed March 30, 2013, *http://www.uspoultry.org.*

10. Penton Media Inc., "Pork Checkoff Surveys Activist Groups' Influence on Children," National Hog Farmer (January 4, 2008), accessed September 15, 2011, *http://nationalhogfarmer.com.*

11. Betsy Booren, "Championing the Beef Industry," accessed April 21, 2012, *http://www.animal.ufl.edu.*

12. Gallup, "Favorability: People in the News," accessed January 29, 2013, *http://www.gallup.com.*

13. Booren, "Championing the Beef Industry."

14. American Meat Institute, "Meat Mythcrushers," accessed October 4, 2011, *http://www.meatmythcrushers.com.*

15. American Meat Institute, "Myth: Americans Eat Too Much Meat and Its Saturated Fat Content Leads to Heart Disease," accessed October 4, 2011, *http://www.meatmythcrushers.com.*

16. US Department of Agriculture, "How Much Food from the Protein Foods Group is Needed Daily?" (2011), accessed April 21, 2012, *http://www.choosemyplate.gov.*

17. Ibid.; US Department of Agriculture Economic Research Service, "Retail Food Commodity Intakes: Mean Amounts of Retail Commodities per Individual, 2001–2002" (2011).

18. *PETA et al. v. Ross et al.,* Superior Court of the State of California, County of Sacramento, Case No. 34-2011-80000886 (2011), "Verified Petition for Writ of Mandate," 9.

19. Ibid., exhibit A, 5–12.

20. Ibid., exhibit A, 6.

21. Ibid., exhibit A, 10.

22. Ibid., exhibit A, 5.

23. *See, for example,* S. A. Bingham et al., "Does Increased Endogenous Formation of N-nitroso Compounds in the Human Colon Explain the Association between Red Meat and Colon Cancer?" *Carcinogenesis* 17, no. 3 (1996): 515–23.

24. Randy Huffman, "Media Needs to Check Background of Pseudo-Medical Animal Rights Group and Cease Coverage of Alarmist and Unscientific

Attack on Meat Products," *American Meat Institute Press Release* (August 1, 2008), accessed September 20, 2011, *http://www.meatami.com.*

25. Stanley Cohen, *States of Denial: Knowing About Atrocities and Suffering* (Cambridge, UK: Polity Press, 2001), 61.

26. J. Patrick Boyle, "Statement of the American Meat Institute on the Petition by Animal Rights and Labor Groups," American Meat Institute Press Release (June 14, 2001), accessed September 20, 2011, *http://www.meatami.com.*

27. Cattlemen's Beef Board and National Cattlemen's Beef Association, "High Quality Protein Promotes Optimal Health," accessed April 24, 2012, *http://www.beefitswhatsfordinner.com.*

28. Cattlemen's Beef Board and National Cattlemen's Beef Association, "Discover the Power of Protein in Lean Beef," accessed April 24, 2012, *http://www.beefitswhatsfordinner.com.*

29. J. E. Morley et al., "Sarcopenia," *The Journal of Laboratory and Clinical Medicine* 137, no. 4 (2001): 231–43, abstract.

30. World Health Organization, Food and Agriculture Organization of the United Nations, and United Nations University, "Protein and Amino Acid Requirements in Human Nutrition" (2007), accessed November 20, 2011, *http://www.who.int.*

31. Ibid., US Department of Agriculture, "What We Eat in America, NHANES 2007–2008," accessed November 15, 2011, *http://www.ars.usda.gov.*

32. US Department of Agriculture, "Content of Selected Protein (g) Foods per Common Measure, Sorted Alphabetically," National Nutrient Database for Standard Reference, Release 24, accessed November 20, 2011, *https://www.ars.usda.gov.*

33. Ibid.

34. Janice Stanger, *The Perfect Formula Diet* (San Diego: Perfect Planet Solutions, 2009), 34.

35. U. D. Register and L. M. Sonnenberg, "The Vegetarian Diet. Scientific and Practical Considerations," *Journal of the American Dietetic Association* 62, no. 3 (1973): 253–61.

36. A 2011 poll by Harris Interactive found that 5 percent of adult Americans are vegetarian and half of these, or 2.5 percent, are vegan. The US Census Bureau advises that the US population is 313.4 million (as of April 25, 2012). The Vegetarian Resource Group, "How Many Adults Are Vegan in the U.S.?" (2011), accessed April 24, 2012, *http://www.vrg.org;* US Census Bureau, "U.S. and World Population Clocks" (2012), accessed April 25, 2012, *http://www.census.gov.*

37. *See, for example,* Campbell and Campbell, *China Study.*

38. National Academy of Sciences, *Dietary Reference Intakes for Energy, Carbohydrate, Fiber, Fat, Fatty Acids, Cholesterol, Protein, and Amino Acids* (Washington, DC: The National Academies Press, 2005): 662.

39. Cattlemen's Beef Board and National Cattlemen's Beef Association, "Powering Up with Protein," accessed April 24, 2012, *http://www.beefitswhatsfordinner.com*.

40. Campbell and Campbell, *China Study*.

41. W. O. Atwater, "Foods: Nutritive Value and Cost," *Farmers' Bulletin* 23 (1894): 18, accessed November 19, 2011, *http://afrsweb.usda.gov*.

42. H. H. Mitchell, "Carl von Voit," *Journal of Nutrition* 13 (1937): 2–13, accessed November 19, 2011, *http://jn.nutrition.org*.

43. Ibid., 9.

44. World Health Organization, "Protein and Amino Acid Requirements," 96.

45. Alice Peloubet Norton, *Food and Dietetics* (Chicago: Home Economics Association, 1907): 229.

Chapter 3

1. Dana Rohrabacher, letter to the author (September 19, 2011).

2. Today's US Senate seat has a $6.5 million price tag, and with elections every six years, incumbents must raise more than $1 million yearly if they hope to get reelected. For those who set their sights a bit lower, running for the US House of Representatives is less expensive—but at $1.1 million, still not cheap. Worse, as members of Congress's lower house must stand for reelection every two years, the pressure to raise money never ends. Members must raise an average of $550,000 annually, or more than three times their annual salary of $174,000, every year they're in office. (Center for Responsive Politics, "Money Wins Presidency and 9 of 10 Congressional Races in Priciest U.S. Election Ever" [2008], accessed July 10, 2012, *http://www.opensecrets.org*.) With 435 representatives, 100 senators and a president, the federal campaign costs add up fast: the 2012 election was the priciest to date at $6 billion. (Center for Responsive Politics, "2012 Election Spending Will Reach $6 Billion, Center for Responsive Politics Predicts" [2012], accessed December 12, 2012, *http://www.opensecrets.org*.) A few candidates in each election choose not to accept campaign contributions, but they usually lose. Just six of forty-nine self-funded candidates for Congress won their elections in 2008. (Center for Responsive Politics, "Money Wins Presidency.")

3. US Senate Office of Public Records, "Lobbying Disclosure Act Database," accessed May 5, 2012, *http://www.senate.gov*.

4. Center for Responsive Politics, "Lobbying: Top Spending 2010," accessed July 10, 2012, *http://www.opensecrets.org*.

5. David J. Wolfson and Mariann Sullivan, "Foxes in the Hen House—Animals, Agribusiness and the Law: A Modern American Fable," in *Animal Rights: Current Debates and New Directions*, Cass R. Sunstein and Martha C. Nussbaum, eds., (New York: Oxford University Press, 2004), 206.

6. D. T. Regan, "Effects of a Favor and Liking on Compliance," *Journal of Experimental Social Psychology* 7 (1971): 627–39.

7. *See, for example,* J. C. Brooks, A. C. Cameron, and C. A. Carter, "Political Action Committee Contributions and U.S. Congressional Voting on Sugar Legislation," *Americana Journal of Agricultural Economics* 80 (1998): 441–54.

8. Public Citizen's Congress Watch, "An Ocean of Milk, a Mountain of Cheese, and a Ton of Money: Contributions from the Dairy PAC to Members of Congress" (1982).

9. Thomas Stratmann, "Can Special Interests Buy Congressional Votes? Evidence from Financial Services Legislation," American Political Science Association 2002 Annual Meeting, Boston, (2002), accessed September 5, 2012, *http://ideas.repec.org.*

10. Marion Nestle, *Food Politics: How the Food Industry Influences Nutrition and Health* (Berkeley: University of California Press, 2007), 105.

11. Rigoberto A. Lopez, "Campaign Contributions and Agricultural Subsidies," *Economics and Politics* 13, no. 3 (2001): 257–78.

12. *See* chapter 2, note 8.

13. Regulations adopted by state and federal agencies also represent a source of law, but because these agencies cannot exceed their statutory mandates, this book's analysis focuses on the underlying statutes.

14. *State v. Rhodes* (1868) 61 N.C. 453, 453

15. *Stephens v. State* (1888) 65 Miss. 329, 331.

16. In 1635 the Irish Parliament prohibited cruelty to sheep and horses; in 1774 the British Parliament prohibited cruelty in the driving of cattle; and the British Martin's Act of 1822 prohibited cruelty to horses, mules, oxen, sheep, and cattle. In the American colonies, the Massachusetts Bay Colony in 1641 prohibited cruelty toward "any brute creatures which are usually kept for man's use." In 1821 Maine banned cruelty to cattle or horses, and in 1829 New York outlawed cruelty to cattle, sheep, and horses.

17. Conn. Gen. Stat. Ann. § 53-247; 1996 Conn. Legis. Serv. P.A. 96-243 (S.H.B. 5801).

18. Thirty-seven states have adopted CFEs. Cody Carlson, "How State Ag-gag Laws Could Stop Animal-Cruelty Whistleblowers," *The Atlantic* (March 25, 2013), accessed April 1, 2013, *http://www.theatlantic.com;* Wolfson and Sullivan, "Foxes in the Hen House," 228, note 20.

18. Erik Marcus, *Meat Market: Animals, Ethics, and Money* (Boston: Brio Press, 2005), 43.

20. National Pork Board, *Swine Care Handbook* (2003), preface, accessed November 3, 2011, *http://www.antwifarms.com.*

21. Sam Howe Verhovek, "Talk of the Town: Burgers v. Oprah," *New York Times* (January 21, 1998).

22. Jia-Rui Chong, "Wood-Chipped Chickens Fuel Outrage," *Los Angeles Times* (November 23, 2003).

23. Kan. Stat. Ann. § 47-1827(c)(4); Mont. Code Ann. § 81-30-103(2)(e); N.D. Cent. Code § 12.1-21.1-02(6); 720 ILCS 215/4(4); Mo. Ann. Stat. § 578.407(3); IA ST § 717A.3A(1)(a); 2012 Utah Laws Ch. 213 (H.B. 187).

24. Kathleen Masterson, "Ag-Gag Law Blows Animal Activists' Cover," *National Public Radio, All Things Considered* (March 10, 2012), accessed May 6, 2012, *http://www.npr.org*.

25. Amanda Radke, "Do You Support Ag-Gag Laws?" *BEEF Daily* (March 14, 2012), accessed May 6, 2012, *http://beefmagazine.com*.

26. Jennifer Jacobs, "Survey Finds Iowa Voters Oppose Prohibiting Secret Animal-Abuse Videos," *Des Moines Register* (March 22, 2011).

27. Erline Aguiluz, "Daniel Clark Charged with Animal Cruelty," *Philadelphia Criminal Law News* (December 13, 2010).

28. Lisa Duchene, "Are Pigs Smarter than Dogs?" Specialty Pet Training, accessed November 3, 2011, *http://www.specialtypettraining.com*.

29. Will Potter, *Green Is the New Red: An Insider's Account of a Social Movement Under Siege* (San Francisco: City Lights Books, 2011).

30. Brendan Greeley, "ALEC's Secrets Revealed; Corporations Flee," *Bloomberg Businessweek* (May 3, 2012).

31. Trust for America's Health, "Supplement to 'F as in Fat: How Obesity Policies are Failing in America, 2007' Obesity-Related Legislation Action in States, Update (2007)," accessed May 4, 2012, *http://healthyamericans.org*.

32. Liza Porteus, Brian Wilson, and the Associated Press, "House Passes 'Cheeseburger Bill'" (March 11, 2004), *FoxNews.com,* accessed May 4, 2012, *http://www.foxnews.com*.

33. Statement of Rep. Pete Stark, *Congressional Record* (October 19, 2005): 23086.

34. *See, for example,* A.R.S. § 13-2301.

35. Dara Lovitz, *Muzzling a Movement: The Effects of Anti-Terrorism Law, Money & Politics on Animal Activism* (Brooklyn: Lantern Books, 2010), 125.

36. Ibid.

37. Potter, *Green Is the New Red,* 135.

38. Jeffrey Record, "Bounding the Global War on Terrorism," *Strategic Studies Institute* (2003): 6, accessed November 10, 2011, *http://www.strategicstudiesinstitute.army.mil*.

39. Geoffrey Nunberg, "Head Games / It All Started with Robespierre / 'Terrorism': The History of a Very Frightening Word," *San Francisco Chronicle* (October 28, 2001).

40. For example, US national security policy defines *terrorism* as "premeditated, politically motivated violence against innocents." Record, "Bounding the Global War on Terrorism," 6.

41. In fact, one of the guiding principles of the Animal Liberation Front, a group which the FBI labels a terrorist organization, is "to take all necessary precautions against harming any animal, human and nonhuman." North American

Animal Liberation Press Office, "Guidelines of the Animal Liberation Front," accessed November 5, 2011, *http://www.animalliberationpressoffice.org.*

42. US Federal Bureau of Investigation, "Major Terrorism Cases: Past and Present" (2011), accessed May 5, 2012, *http://www.fbi.gov.*

43. Mike German, *Thinking Like a Terrorist* (Washington, DC: Potomac Books, 2007), 154–55.

44. *See* chapter 2, note 9.

45. Congress has passed such important legislation as the Civil Rights Act (outlawing discrimination in public places, schools, and employment), the Voting Rights Act (outlawing discriminatory practices used to deny blacks the right to vote), and the Americans with Disabilities Act (outlawing discrimination against those with disabilities). Federal courts have also played a key role in protecting the downtrodden, enforcing unpopular laws in regions where state courts could not, and occasionally issuing an important decision which interprets existing law to provide new protections—such as the US Supreme Court's 1954 decision in *Brown v. Board of Education*, declaring racial segregation in public schools unconstitutional. The federal executive branch has also played a role in extending important social change through executive order—such as Abraham Lincoln's freeing southern slaves through the Emancipation Proclamation, or Harry Truman's order prohibiting racial or religious discrimination in the armed services.

46. For example, in 2010, legislators in New Mexico received no salary, while those in Texas and South Dakota received salaries of $7,200 and $6,000 per year, respectively. National Conference of State Legislatures, "2010 Legislator Compensation Data," accessed November 2, 2011, *http://www.ncsl.org.*

47. The European Union limits transport times differently depending on the animal involved and its age: unweaned animals (those still drinking milk) are limited to nine hours of travel; cattle, sheep, and goats are limited to fourteen hours of travel; pigs are limited to twenty-four hours of travel, provided they have continuous access to water; and horses are limited to twenty-four hours of travel, provided they have access to water every eight hours. European Council Regulation No. 1/2005, "Protection of Animals During Transport and Related Operations and Amending Directives" (December 22, 2004).

48. 49 U.S.C. § 80502(a)(2)(A).

49. 49 U.S.C. § 80502(a)(2)(B).

50. *See, for example, U.S. v. Lehigh Val R Co* (3d Cir. 1913) 204 F. 705, 708: "The lambs were not fed within the statutory period; but we repeat that in our opinion the fact is accounted for by negligence, and cannot be attributed to a 'knowing and willful' disregard of duty."

51. Animal Welfare Institute, "Legal Protections for Farm Animals During Transport" (August 2010), accessed November 2, 2011, *http://www.awionline.org.*

52. Temple Grandin, *Thinking in Pictures: My Life with Autism* (New York: First Vintage Books, 1996), 153–54.

53. Humane Farming Association et al., *Petition for Enforcement of Humane Slaughter and Animal Cruelty Laws* (2000), affidavit #3, accessed November 3, 2011, *http://www.citizen.org.*

54. "Groups Petition USDA to Enforce Humane Slaughter Act," *Public Citizen* (2001), accessed November 3, 2011, *http://www.citizen.org.*

55. Farm Security and Rural Investment Act of 2002 § 10305.

56. Stanley Painter, "Testimony Before the Domestic Policy Subcommittee of the House Committee on Oversight and Government Reform" (2008), accessed November 2, 2011, *http://njcfil.com.*

57. Matthew Madia, "Federal Meat Inspectors Spread Thin as Recalls Rise," *OMB Watch* (2007), accessed November 5, 2011, *http://ombwatch.org.*

58. American Meat Institute, "American Meat Institute Responds to Questions about Industry Support of Federal Oversight" (2008), accessed November 2, 2011, *http://www.meatami.com.*

59. Nathan Runkle, email to author (December 13, 2011).

60. Painter, "Testimony Before the Domestic Policy Subcommittee."

61. US Department of Agriculture, "USDA Announces Measures to Improve Humane Handling Enforcement," Press Release (2010), accessed November 2, 2011, *http://www.fsis.usda.gov.* The new measures included:

    • Requiring that nonambulatory cattle be promptly and humanely euthanized.

    • Becoming more responsive to petitions from animal protection groups.

    • Providing meat inspectors with better training in humane animal handling.

    • Appointing an ombudsman to whom inspectors can voice complaints outside of the standard reporting structure.

    • Auditing appeals of inspectors' reports of humane handling violations to determine whether, as claimed by inspectors, supervisors routinely dismiss violation reports at the request of the meat industry.

62. Clarence P. Dresser, "Vanderbilt in the West," *New York Times* (October 9, 1882).

63. Robert B. Reich, *Supercapitalism: The Transformation of Business, Democracy, and Everyday Life* (New York: Vintage Books, 2007), 214.

Chapter 4

1. AquaBounty Technologies, "Myths and Facts," accessed May 7, 2012, *http://www.aquabounty.com.*

2. "Would You Eat Genetically-Modified Salmon if Approved by the FDA?" poll, *Wall Street Journal* (2009), accessed May 10, 2012, *http://online.wsj.com.*

3. American Academy of Environmental Medicine, "Genetically Modified Foods," position paper (2009), accessed November 4, 2011, *http://herbogeminis.com*.

4. *See, for example*, Chelsea Snell et al., "Assessment of the Health Impact of GM Plant Diets in Long-Term and Multigenerational Animal Feeding Trials: A Literature Review," *Food and Chemical Toxicology* 40, nos. 3–4 (2012): 1134–48.

5. *See, for example*, M. Schrøder, "A 90-Day Safety Study of Genetically Modified Rice Expressing Cry1AB Protein (*Bacillus thuringiensis* toxin) in Wistar Rats," *Food and Chemical Toxicology* 45, no. 3 (2007): 339–49.

6. Jeffrey Smith, *Genetic Roulette: The Documented Health Risks of Genetically Engineered Foods* (White River Junction, Vermont: Chelsea Green, 2007).

7. US Food and Drug Administration, "Guidance for Industry: Voluntary Labeling Indicating Whether Foods Have or Have Not Been Developed Using Bioengineering—Draft Guidance" (2000), accessed May 7, 2012, *http://www.fda.gov*.

8. Keith R. Schneider and Renee Goodrich Schneider, "Genetically Modified Food," University of Florida IFAS Extension FSHN02-2 (2002), accessed May 10, 2012, *http://edis.ifas.ufl.edu*.

9. Monica Eng, "FDA Finally Responds to GMO-Labeling Campaign but Differs on Numbers of Supporters," *Chicago Tribune* (March 28, 2012), accessed May 7, 2012, *http://www.chicagotribune.com*.

10. Ibid.

11. George Stigler, "The Theory of Economic Regulation," *Bell Journal of Economics and Management Science* 3 (1971): 3–18.

12. 59 Federal Register 6280 (1994).

13. *Int'l Dairy Foods Ass'n v. Boggs* (6th Cir. 2010) 622 F.3d 628, 636 (emphasis added).

14. Ibid.

15. The FDA's concern apparently stemmed from a 1968 report, known as the Swann Report, prepared by four UK governmental branches and submitted to Parliament. The report found:

> A dramatic increase over the years in the numbers of strains of enteric bacteria of animal origin which show resistance to one or more antibiotics. Further, these resistant strains are able to transmit this resistance to other bacteria. This resistance has resulted from the use of antibiotics for growth promotion and other purposes in farm livestock. . . . There is ample and incontrovertible evidence to show that man may commonly ingest enteric bacteria of animal origin.

   M. M. Swann, K. L. Baxter, and H. I. Field, "1969 Report of the Joint Committee on the Use of Antibiotics in Animal Husbandry and Veterinary Medicine," Her Majesty's Stationery Office (1969), 60.

16. 37 Federal Register 2,444 (February 1, 1972).

17. 21 U.S.C. § 360b(e)(1).

18. 42 Federal Register 43,772 (August 30, 1977), 43,792; 42 Federal Register 56,264 (October 21, 1977), 56,288.

19. Greg Cima, "FDA Cancels 1977 Drug Withdrawal Bids," *Journal of the American Veterinary Medical Association News* (February 15, 2012), accessed May 10, 2012, *http://www.avma.org.*

20. After summarizing decades of clinical research into animal drug use, the report noted: "The scientific community generally agrees that antimicrobial drug use is a key driver for the emergence of antimicrobial-resistant bacteria." US Food and Drug Administration, "The Judicious Use of Medically Important Antimicrobial Drugs in Food-Producing Animals," *Guidance for Industry* 209 (2012), accessed May 10, 2012, *http://www.fda.gov.*

21. House of Representatives Reports No. 95-1290 (1978); House of Representatives Reports No. 96-1095 (1980); Senate Reports No. 97-248 (1981).

22. *Natural Resources Defense Council et al. v. U.S. Food and Drug Administration et al.,* No. 11 Civ. 3562 (S.D.N.Y. Mar. 22, 2012)

23. US Food and Drug Administration, "The Judicious Use of Medically Important Antimicrobial Drugs," 13.

24. USDA National Institute of Food and Agriculture, "Markets, Trade & Policy Overview," accessed March 12, 2011, *http://www.nifa.usda.gov.*

25. Jeff Herman, "Saving U.S. Dietary Advice from Conflicts of Interest," *Food & Drug Law Journal* 65 (2010): 285–326.

26. Harvard School of Public Health, "Food Pyramids and Plates: What Should You Really Eat?" accessed November 3, 2011, *http://www.hsph.harvard.edu.*

27. US Department of Agriculture and US Department of Health and Human Services, "2010 Dietary Guidelines for Americans" (2010): x, accessed November 4, 2011, *http://www.cnpp.usda.gov.*

28. Ibid., xi.

29. *Physicians' Committee for Responsible Medicine v. Vilsack et al.,* No. 1:11-cv-00038-RJL (2011 U.S. Dist. D.C. Cir.).

30. 7 U.S.C. § 5341(a).

31. US Department of Agriculture, "Your Personal Path to Health: Steps to a Healthier You!" accessed November 5, 2011, *www.choosemyplate.gov.*

32. Michael Moss, "While Warning about Fat, U.S. Pushes Cheese Sales," *New York Times* (November 6, 2010).

33. Ibid.

34. Becker, "Federal Farm Promotion ('Check-Off') Programs."

35. Kim Krisberg, "Dietary Guidelines, Food Pyramid Facing Scrutiny: Officials to Update Recommendations," *The Nation's Health* 33, no. 9 (2003).

36. Phyllis K. Fong, "Food Safety and Inspection Service Sampling and Testing for *E. coli*" USDA, memorandum to Deputy Secretary Charles F. Conner (2008).

37. Fong, "Food Safety," 3.
38. Office of Inspector General, "Audit Report 24601-0007-KC," USDA (2008), iii.
39. Office of Inspector General, "Audit Report 24601-08-KC," USDA (2010).
40. Kimberly Kindy and Lyndsey Layton, "Purity of Federal 'Organic' Label Is Questioned," *The Washington Post* (July 3, 2009).
41. Ibid.
42. US Department of Agriculture, "Know Your Farmer, Know Your Food," accessed May 5, 2011, *http://www.usda.gov.*
43. Robert Kenner, director, *Food, Inc.* (Participant Media, 2008).
44. Clarence Thomas (US Supreme Court justice since 1991) was an attorney for Monsanto from 1976 to 1979; Donald Rumsfeld (US secretary of defense from 1975 to 1977 and 2001 to 2006) was CEO from 1977 to 1985 of G. D. Searle & Co., later acquired by Monsanto; Mickey Kantor (US trade representative from 1993 to 1997 and US secretary of commerce from 1996 to 1997) was formerly on the Monsanto board of directors; Margaret Miller (FDA branch chief during the 1990s) was a Monsanto chemical lab supervisor from 1985 to 1989; Michael Taylor (FDA deputy commissioner for policy from 1991 to 1994 and deputy commissioner for foods since 2010) was a lawyer for Monsanto from 1989 to 1991 and Monsanto's vice president for public policy from 1998 to 2000; Linda Fisher (EPA deputy administrator from 2001 to 2003) was Monsanto's vice president for public affairs from 1995 to 2000.
45. Philip Mattera, "USDA Inc.: How Agribusiness Has Hijacked Regulatory Policy at the U.S. Department of Agriculture," *Food and Agriculture Conference of the Organization for Competitive Markets* (2004), accessed October 25, 2011, *http://www.nffc.net.*
46. Ibid., 10–11. Specifically, the report found that:

- USDA Secretary Ann M. Veneman served on the board of biotech company Calgene (later acquired by Monsanto).

- Veneman's Chief of Staff Dale Moore was executive director for legislative affairs of the National Cattlemen's Beef Association (NCBA).

- Veneman's Deputy Chief of Staff Michael Torrery was a vice president at the International Dairy Foods Association.

- Director of Communications Alisa Harrison was executive director of public relations at NCBA.

- Deputy Secretary James Moseley was a partner in Infinity Pork LLC, a factory farm operator in Indiana.

- Undersecretary J. B. Penn was an executive of Sparks Companies, an agribusiness consulting firm.

- Undersecretary Elsa Murano conducted industry-sponsored research while a university professor.

- Undersecretary Joseph Jen was director of research at Campbell Soup Company's Campbell Institute of Research and Technology.

- Deputy Undersecretary Floyd D. Gaibler was executive director of the National Cheese Institute and the American Butter Institute, which are funded by the dairy industry.

- Deputy Undersecretary Kate Coler was director of government relations for the Food Marketing Institute.

- Deputy Undersecretary Charles Lambert spent fifteen years working for NCBA.

- Assistant Secretary for Congressional Relations Mary Waters was a senior director and legislative counsel for ConAgra Foods.

47. Eric Schlosser, "The Cow Jumped Over the U.S.D.A.," *New York Times* (January 2, 2004).

48. Organic Consumers Association, "Six Reasons Why Obama Appointing Monsanto's Buddy, Former Iowa Governor Vilsack, for USDA Head Would Be a Terrible Idea" (2008), accessed September 6, 2012, *http://www.organic-consumers.org*.

49. Organic Consumers Association, "Six Reasons"; John Robbins, *No Happy Cows: Dispatches from the Frontlines of the Food Revolution* (San Francisco: Conari Press, 2012), xi.

50. Tom Philpott, "In a Stunning Reversal, USDA Chief Vilsack Greenlights Monsanto's Alfalfa," *Grist* (2011), accessed September 6, 2012, *http://grist.org*.

51. Rory Freedman and Kim Barnouin, *Skinny Bitch* (Philadelphia: Running Press, 2005), 92.

Chapter 5

1. US Bureau of Labor Statistics, "Consumer Price Index Average Price Data," accessed September 1, 2011, *http://www.bls.gov*; Brian W. Gould, "Understanding Dairy Markets," online database, accessed September 2, 2011, *http://future.aae.wisc.edu*.

2. David Leonhardt, "What's Wrong with This Chart?" *New York Times* (May 20, 2009).

3. Erik Marcus, *Meat Market: Animals, Ethics & Money* (Boston: Brio Press, 2005).

4. Michael Roberts, "U.S. Animal Agriculture: Making the Case for Productivity," *AgBioForum* 3 (2000).

5. Ibid.

6. US Centers for Disease Control and Prevention, *Sustaining State Programs for Tobacco Control: Data Highlights 2006* (Atlanta: US Department of

Health and Human Services, 2006), accessed October 26, 2011, *http://www.cdc.gov.*

7. National Research Council of the National Academies, *The Hidden Costs of Energy: Unpriced Consequences of Energy Production and Use* (Washington, DC: National Academies Press, 2010), accessed May 19, 2012, *http://www.nap.edu.*

8. This method isn't completely precise because not all consumers incur all costs. However, most consumers do incur most costs, so it provides a reasonable high-level picture, and the alternative methods of calculation are unnecessarily complex.

9. The annual retail sales figure of $238 billion in 2010 dollars, $251 billion today, is the sum of (a) for food consumed at home, total consumer unit expenditures on animal foods and (b) for food consumed away from home, the ratio of (i) spending (by unit type) on animal foods to (ii) total spending on food consumed at home, multiplied by (iii) total spending on food consumed away from home. This calculation shows that the portion of retail food dollars spent on animal foods is about 32 percent. US Bureau of Labor Statistics, "Composition of Consumer Unit," *Consumer Expenditure Survey* (2011), accessed December 3, 2011, *http://www.bls.gov.*

10. For details, *see* Appendix B.

11. The exact figure might be slightly more or less than $665 billion, since producers might absorb some of the cost increase themselves, or they might pass along a greater increase to consumers.

12. Fruits and vegetables, for example, generate few costs associated with antibiotics, hazardous wastes, global warming, or real estate devaluation.

13. Physicians Committee for Responsible Medicine, "Agriculture and Health Policies in Conflict" (2011), accessed October 23, 2011, *http://www.pcrm.org.*

14. Bruce Sundquist, "Economics, Politics and History of Irrigation" (2010), accessed October 22, 2011, *http://home.windstream.net.*

15. Physicians Committee for Responsible Medicine, "Agriculture and Health Policies in Conflict."

16. $28.9 billion of this total is from the USDA's 2013 budget: Farm Service Agency ($12.1 billion); Risk Management Agency ($9.6 billion); Research, Education, and Economics ($2.7 billion); Marketing and Regulatory Programs ($2.4 billion); and Foreign Agricultural Service ($2.1 billion). US Department of Agriculture, "Budget for 2013," accessed September 28, 2012, *http://www.usda.gov.* The remaining $1.9 billion—adjusted from $1.8 billion in 2009 dollars—of the total is federal irrigation subsidies reported in Grey, Clark, Shih and Associates Limited, "Farming the Mailbox: U.S. Federal and State Subsidies to Agriculture—Study Prepared for Dairy Farmers of Canada" (2010), accessed August 8, 2012, *http://www.greyclark.com.*

17. Grey et al. report $21.5 billion in state irrigation spending and another $3.2 billion in general state and local subsidies (both in 2009 dollars), for

an inflation-adjusted total of $26.5 billion. Grey, Clark, Shih and Associates Limited, "Farming the Mailbox."

18. Unfortunately, this report does not show its math. (Physicians Committee for Responsible Medicine, "Agriculture and Health Policies in Conflict.") However, PCRM's figure is consistent, in a general way, with agricultural land-use statistics (assuming such statistics represent a reasonable gauge for how subsidy dollars are spent). Thus, according to the USDA, 1,247 million acres of US land are used for all agricultural purposes, and 82 percent of this total, or 1,028 million acres, is dedicated to raising livestock (i.e., to graze animals or grow feed crops). (Ruben N. Lubowski et al., "Major Uses of Land in the United States, 2002," USDA Economic Research Service (2006), accessed August 19, 2012, *http://www.ers.usda.gov.*)

19. Annual US fishing subsidies total $1.8 billion in 2003 dollars, or $2.3 billion today. U. Rashid Sumaila et al., "A Bottom-Up Re-Estimation of Global Fisheries Subsidies," *Journal of Bioeconomics* 12 (2010): 201–25.

20. Environmental Working Group, "Farm Subsidy Database," accessed October 22, 2011, *http://farm.ewg.org.*

21. US Department of Agriculture, "Feed Grains Data Delivery System," accessed October 23, 2011, *http://www.ers.usda.gov.*

22. Ibid.

23. Timothy Wise, "Identifying the Real Winners from U.S. Agricultural Policies" (working paper, 05-07, Global Development and Environment Institute 2005), accessed October 23, 2011, *http://ase.tufts.edu.*

24. Elanor Starmer and Timothy Wise, "Feeding at the Trough: Industrial Livestock Firms Saved $35 Billion from Low Feed Prices" (policy brief, 07-03, Global Development and Environment Institute Tufts University 2007), accessed October 23, 2011, *http://www.ase.tufts.edu.*

25. Timothy Egan, "Failing Farmers Learn to Profit from Federal Aid," *New York Times* (December 24, 2000).

26. Wise, "Identifying the Real Winners."

27. Carol A. Jones, Hisham El-Osta, and Robert Green, "Economic Well-Being of Farm Households," USDA Economic Research Service (2006), accessed October 22, 2011, *http://www.ers.usda.gov.*

28. Ibid.

29. Thomas A. Fogarty, "Freedom to Farm? Not Likely?" *USA Today* (January 2, 2002).

30. Grunwald, "Why Our Farm Policy Is Failing."

31. Jan L. Flora et al., "Social and Community Impacts," in *Iowa Concentrated Animal Feeding Operations Air Quality Study,* ed. Iowa State University et al. (Iowa City: University of Iowa Printing Service, 2002), 147.

32. Grunwald, "Why Our Farm Policy Is Failing."

33. Elanor Starmer and Timothy Wise, "Living High on the Hog: Factory Farms, Federal Policy, and the Structural Transformation of Swine Production"

(working paper, 07-04, Global Development and Environment Institute 2007), accessed October 23, 2011, *http://ase.tufts.edu.*

34. Mary Hendrickson and William Heffernan, "Concentration of Agricultural Markets," Food Circles Networking Project (2005), accessed October 23, 2011, *http://www.foodcircles.missouri.edu.*

35. James Kliebenstein, "Economic and Associated Social and Environmental Issues with Large Scale Livestock Production Systems" (concept paper for *World Bank Workshop on Sustainable Intensification of Agricultural Systems: Linking Policy, Institutions and Technology,* Ames, Iowa: 1998).

36. Jim Jacobson and Chris Bedford, *The Manure Money Pit: How Environmental Tax Subsidies to Hog Confinements Impact Iowa's Counties* (Des Moines: Human Society of the United States, 2003).

37. International Labor Rights Forum, "NAFTA, Creating a Sweatfree World: Changing Global Trading Rules," accessed October 22, 2011, *http://www.laborrights.org.*

38. Max Borders and H. Sterling Burnett, "Farm Subsidies: Devastating the World's Poor and the Environment," National Center for Policy Analysis (2006), accessed October 23, 2011, *http://www.ncpa.org.*

39. Ibid., 1.

40. Charles F. Conner, "Conference Call with Reporters: Announcement of a New Farm Bill from Congress," USDA Transcript (May 9, 2008).

41. Rigoberto A. Lopez, "Campaign Contributions and Agricultural Subsidies," *Economics and Politics* 13, no. 3 (2001): 257–78.

42. Elanor Starmer, Aimee Witteman, and Timothy A. Wise, "Feeding the Factory Farm: Implicit Subsidies in the Broiler Chicken Industry" (working paper, 06-03, Global Development and Environment Institute 2006), 5, accessed October 23, 2011, *http://www.ase.tufts.edu.*

43. Incidentally, there is some disagreement among commentators over whether certain subsidies, known as decoupled payments because they are made regardless of production levels or commodity prices, influence farmers. If these subsidies don't actually influence farmers' decisions to grow feed crops, then they wouldn't keep feed crop prices low, which means they wouldn't help livestock producers in the ways described. For example, some research suggests that subsidy payments that are fully or partially decoupled from production, such as those paid solely based on land's historic use, have only a modest effect on crop production. (William Lin and Robert Dismukes, "Supply Response under Risk: Implications for Counter-Cyclical Payments' Production Impact," *Review of Agricultural Economics* 29, no. 1 [2006]: 64–86; Barry K. Goodwin and Ashok K. Mishra, "Are 'Decoupled' Farm Program Payments Really Decoupled? An Empirical Evaluation," *American Journal of Agricultural Economics* 88, no. 1 [2006]: 73–89.) As decoupled payments are thought to have little impact on global trade and are thus favored by

the World Trade Organization, the United States has increasingly sought to provide its farmers with decoupled payments in the past couple of decades.

This line of logic might suggest that the subsidy tail isn't actually wagging the feed-producing dog. Nonetheless, there is evidence that a significant portion of farm subsidies *does* influence farmers who grow feed crops, which means these subsidies *do* benefit livestock producers. For example, one study found that more than two-thirds of farmers spend decoupled payments on farm purposes like operating costs, capital expenditures, and farm debt, and larger operators are particularly likely to spend payments on the farm. (Barry K. Goodwin and Ashok K. Mishra, "Another Look at Decoupling: Additional Evidence of the Production Effects of Direct Payments," *American Journal of Agricultural Economics* 87, no. 5 [2005]: 1200–10.) As a result, the study's authors found that decoupled subsidy payments "have important effects on production" (Ibid., 1206). Moreover, the latest farm bill seeks to continue both coupled and decoupled subsidies, and there's no doubt that coupled measures like price supports have a direct effect on production. Some critics even argue that the more than $80 billion spent yearly on food stamps is little more than an enormous price support program, which made the decision to omit that program from the subsidy calculation, as too attenuated, a difficult one. (Grey, Clark, Shih and Associates Limited, "Farming the Mailbox," 256.) Thus, not only does it seem that decoupled payments do in fact influence production decisions, but larger operators, who get most of the subsidy funds, are the most likely to be so influenced.

44. Extra equity value is estimated as follows: the US meat industry's market capitalization is $38 billion or 17.2 times earnings; subsidies to the industry reduce operating costs by roughly 10 percent (the midpoint of the range of 5 to 15 percent found by Starmer and Wise); the effect of reducing operating costs by 10 percent is to increase earnings by 10 percent and hence (because market capitalization is based on earnings) to increase market capitalization by 10 percent (thus, without subsidies, market capitalization would be $38 billion – ($38 billion x 0.10) or about $34.2 billion). Annual dividend payments are calculated by multiplying total industry market capitalization of $38 billion by average industry dividend yield rate of 0.6 percent. Because net profit margin is only 2.9 percent but subsidies reduce operating expenses by roughly 10 percent, without subsidies these companies would arguably have no cash with which to pay dividends. Yahoo! Finance, "Meat Products: Industry Statistics," accessed October 27, 2012, *http://biz.yahoo.com*; Starmer and Wise, "Feeding at the Trough."

45. The subsidy figure is 13 percent of Tyson's FY 2010 operating expenses, the amount by which researchers estimate federal crop subsidies help lower Tyson's costs. Starmer and Wise, "Feeding at the Trough"; salary and expense data from US Securities and Exchange Commission, "Filings and Forms," accessed October 23, 2011, *http://www.sec.gov*.

46. David Herszenhorn, "Reaching Well Beyond the Farm," *New York Times* (May 20, 2008).
47. David Brooks, "Talking Versus Doing," *New York Times* (May 20, 2008).
48. Quoted in "A Mediocre Farm Bill," *New York Times* (June 24, 2012).
49. Steve Forbes, "Railroading the Taxpayer," *Fact and Comment,* Forbes.com (August 11, 2010).
50. Starmer, Witteman, and Wise, "Feeding the Factory Farm," 32.
51. See Appendix C, table C2.

Chapter 6

1. Véronique L. Roger et al., "Heart Disease and Stroke Statistics—2011 Update," *Circulation* 123 (2011): e18–e209.
2. One in twenty-five estimate based on 2008 US population of 305 million. Department of Health and Human Services, "National Diabetes Statistics," National Diabetes Information Clearinghouse (2011), accessed January 14, 2012, *http://diabetes.niddk.nih.gov;* American Cancer Society, "Cancer Prevalence: How Many People Have Cancer?" (2008), accessed January 14, 2012, *http://www.cancer.org.*
3. World Health Organization, "Obesity and Overweight," accessed December 25, 2011, *http://www.who.int.*
4. Pew Commission on Industrial Farm Animal Production, "Putting Meat on the Table: Industrial Farm Animal Production in America" (2008), accessed October 25, 2012, *http://www.ncifap.org.*
5. US Department of Agriculture, "Nutrient Intakes from Food: Mean Amounts Consumed per Individual, One Day 2005–2006," accessed November 15, 2011, *http://www.ars.usda.gov;* Department of Health and Human Services, "Know Your Fats," accessed January 1, 2012, *http://www.csrees.usda.gov.*
6. The Office of Chief Medical Examiner, "Report of External Examination: Robert Atkins," The City of New York (2003), accessed November 15, 2011, *http://www.thesmokinggun.com.*
7. *See, for example,* Rashmi Sinha et al., "Meat Intake and Mortality: A Prospective Study of Over Half a Million People," *Archives of Internal Medicine* 169, no. 6 (2009): 562–71; Teresa T. Fung et al., "Prospective Study of Major Dietary Patterns and Stroke Risk in Women," *Stroke* 35 (2004): 2014–19; Walker, "Diet in the Prevention of Cancer"; *Iqbal et al.,* "Dietary Patterns and Acute Myocardial Infarction."
8. M. D. Kontogianni et al., "Relationship between Meat Intake and the Development of Acute Coronary Syndromes: The CARDIO2000 Case–Control Study," *European Journal of Clinical Nutrition* 62 (2008): 171–77.
9. US Department of Agriculture, "Poultry Supply and Disappearance" (2011), accessed May 23, 2012, *http://www.ers.usda.gov.*
10. Campbell and Campbell, *China Study.*

11. National Academy of Sciences, *Dietary Reference Intakes for Energy, Carbohydrate, Fiber, Fat, Fatty Acids, Cholesterol, Protein, and Amino Acids* (Washington, DC: The National Academies Press, 2005), 103.

12. US Department of Agriculture, "Cholesterol (mg) Content of Selected Foods per Common Measure, Sorted by Nutrient Content," National Nutrient Database for Standard Reference, Release 21 (2008).

13. An Pan et al., "Red Meat Consumption and Risk of Type 2 Diabetes: 3 Cohorts of U.S. Adults and An Updated Meta-Analysis," *American Journal of Clinical Nutrition* 94, no. 4 (2011): 1088–96, abstract.

14. T. Colin Campbell's research found that eating just 7 grams of meat per day increased human subjects' risk of cancer. Campbell and Campbell, *China Study.*

15. *See, for example,* Gary E. Fraser, "Associations between Diet and Cancer, Ischemic Heart Disease, and All-Cause Mortality in Non-Hispanic White California Seventh-Day Adventists," *American Journal of Clinical Nutrition* 70, no. 3 (1999): 532s–38s; Rashmi Sinha et al., "Meat and Meat-Related Compounds and Risk of Prostate Cancer in a Large Prospective Cohort Study in the United States," *American Journal of Epidemiology* 170, no. 9 (2009): 1165–77; Tanya Agurs-Collins et al., "Dietary Patterns and Breast Cancer Risk in Women Participating in the Black Women's Health Study," *American Journal of Clinical Nutrition* 90 (2009): 621–28; Eleni Linos et al., "Red Meat Consumption During Adolescence among Premenopausal Women and Risk of Breast Cancer," *Cancer Epidemiology, Biomarkers & Prevention* 17 (2008) 2146–51; Ann Chao et al., "Meat Consumption and Risk of Colorectal Cancer," *Journal of the American Medical Association* 293, no. 2 (2005): 172–82.

16. Murray Waldman and Marjorie Lamb, *Dying for a Hamburger: Modern Meat Processing and the Epidemic of Alzheimer's Disease* (New York: St. Martin's Press, 2004).

17. C. G. Coimbra and V. B. C. Junqueira, "High Doses of Riboflavin and the Elimination of Dietary Red Meat Promote the Recovery of Some Motor Functions in Parkinson's Disease Patients," *Brazilian Journal of Medical and Biological Research* 36, no. 10 (2003): 1409–17.

18. Ibrahim Abubakar et al., "A Case-Control Study of Drinking Water and Dairy Products in Crohn's Disease—Further Investigation of the Possible Role of Mycobacterium Avium Paratuberculosis," *American Journal of Epidemiology* 165, no. 7 (2007): 776–83.

19. Hyon K. Choi, "Diet and Rheumatoid Arthritis: Red Meat and Beyond," *Arthritis & Rheumatism,* 50 (2004): 3745–47.

20. Hyon K. Choi et al., "Purine-Rich Foods, Dairy and Protein Intake, and the Risk of Gout in Men," *New England Journal of Medicine* 350 (2004): 1093–1103.

21. Paul N. Appleby, Naomi E. Allen, and Timothy J. Key, "Diet, Vegetarianism, and Cataract Risk," *American Journal of Clinical Nutrition* 93, no. 5 (2011): 1128–35.

22. ChartsBin, "Current Annual Worldwide Meat Consumption Per Capita," accessed December 27, 2011, *http://chartsbin.com.*

23. US Centers for Disease Control and Prevention, "U.S. Obesity Trends," accessed December 27, 2011, *http://www.cdc.gov;* World Cancer Research Fund International, "Data Comparing More and Less Developed Countries," accessed December 27, 2011, *http://www.wcrf.org;* American Cancer Society, "Cancer Facts and Figures 2011," accessed December 27, 2011, *http://www. cancer.org;* National Cancer Institute, "Surveillance Epidemiology and End Results," accessed December 27, 2011, *http://seer.cancer.gov;* World Diabetes Foundation, "Diabetes Facts," accessed December 27, 2011, *http://www. worlddiabetesfoundation.org;* American Diabetes Association, "Diabetes Statistics," accessed December 27, 2011, *http://www.diabetes.org.*

24. US Central Intelligence Agency, "Life Expectancy at Birth," The World Factbook, accessed January 1, 2012, *https://www.cia.gov.*

25. Gary E. Fraser and David J. Shavlik, "Ten Years of Life—Is It a Matter of Choice?" *Archives of Internal Medicine* 161 (2001): 1645–52; Kontogianni et al., "Meat Intake and Acute Coronary Syndromes"; Serena Tonstad et al., "Type of Vegetarian Diet, Body Weight and Prevalence of Type 2 Diabetes," *Diabetes Care* 32 (2009): 791–96; Ann Chao et al., "Meat Consumption and Risk of Colorectal Cancer," *Journal of the American Medical Association* 293, no. 2 (2005): 172–82.

26. P. H. Gann et al., "Prospective Study of Plasma Fatty Acids and Risk of Prostate Cancer," *Journal of the National Cancer Institute* 86, no. 4 (1994): 281–86; Eunyoung Cho, Wendy Y. Chen, and David J. Hunter, "Red Meat Intake and Risk of Breast Cancer Among Premenopausal Women," *Archives of Internal Medicine* 166 no. 20 (2006): 2253–59.

27. D. A. Snowdon, R. L. Phillips, and G. E. Fraser, "Meat Consumption and Fatal Ischemic Heart Disease," *Preventative Medicine* 13, no. 5 (1984): 490–500; J. Chang-Claude, R. Frentzel-Beyme, and U. Eilber, "Mortality Pattern of German Vegetarians After 11 Years of Follow-Up," *Epidemiology* 3, no. 5 (1992): 395–401.

28. Walker, "Diet in the Prevention of Cancer"; Iqbal et al., "Dietary Patterns and Acute Myocardial Infarction."

29. Hana Ross and Frank J. Chaloupka, "The Effect of Cigarette Prices on Youth Smoking," *Health Economics* 12, no. 3 (2003): 217–30.

30. Prices adjusted for inflation. US Census Bureau, "US Statistical Abstract" (1940), accessed February 8, 2012, *http://www.census.gov;* US Bureau of Labor Statistics, "Average Prices" (2011), accessed February 8, 2012, *http:// www.bls.gov.*

31. Tatiana Andreyeva, Michael W. Long, and Kelly D. Brownell, "The Impact of Food Prices on Consumption: A Systematic Review of Research on the Price Elasticity of Demand for Food," *American Journal of Public Health* 100, no. 2 (2010): 216–22.

32. Ronald M. Ayers and Robert A. Collinge, *Microeconomics: Explore & Apply* (New Jersey: Prentice Hall, 2003), 120.

33. Ibid.

34. Technically speaking, except where conspicuous consumption is at work (meaning an increase in price increases the quantity demanded), elasticity figures are given in negative numbers. So the actual price elasticity of demand for dairy is -0.65. However, for convenience, this book follows the popular convention of using absolute values rather than negative values for elasticity numbers.

35. Ayers and Collinge, *Microeconomics.*

36. This calculation is shown in Appendix C, table C1.

37. Rivera-Ferre, "A Chicken and Egg Paradigm?" 103 (emphasis added).

38. Ibid.

39. Ibid., 102.

40. 170 x 0.65 = 1.1 or 110 percent. Since demand cannot fall by more than 100 percent, this result is reduced to 100 percent.

41. National Academy of Sciences, "Dietary Reference Intakes," 425, 549.

42. Ibid., 441, 542 (emphasis added).

43. Ibid., 13.

44. Ibid., 103.

45. European Food Safety Authority, "Dietary Reference Values for Fats."

46. Several of these studies are explained in more detail in chapter 10.

47. Food and Drug Administration, "Guidance for Industry, a Food Labeling Guide" (2011), accessed May 20, 2012 at *http://www.fda.gov;* USDA Economic Research Service, "Retail Food Commodity Intakes: Mean Amounts of Retail Commodities per Individual, 2001–2002" (2011).

48. Litjen Tan, "Diagnosis and Management of Foodborne Illnesses: A Primer for Physicians and Other Health Care Professionals," *Morbidity and Mortality Weekly Report* 53 (2004): 1–33.

49. Paul S. Mead et al., "Food-Related Illness and Death in the United States," *Emerging Infectious Diseases* 5, no. 5 (1999), accessed November 15, 2011, *http://wwwnc.cdc.gov.*

50. Eric Schlosser, *Fast Food Nation: The Dark Side of the All-American Meal* (New York: Houghton Mifflin, 2002), 197.

51. *Consumer Reports*, "Dirty Birds: Even 'Premium' Chickens Harbor Dangerous Bacteria," January 2007, accessed November 15, 2011, *http://www.usapeec.org.*

52. Ibid.

53. USDA Food Safety and Inspection Service, "Nationwide Federal Plant Raw Ground Beef Microbiological Survey" (1996), accessed November 15, 2011, *http://www.fsis.usda.gov.*

54. US Department of Health and Human Services, Centers for Disease Control and Prevention, Food and Drug Administration, and US Department of Agriculture, "National Antimicrobial Resistance Monitoring System

(NARMS) 2009 Executive Report" (2009), 92, accessed November 20, 2011, *http://www.fda.gov.*

55. Ibid.

56. Associated Press, "Source of Tainted Spinach Finally Pinpointed," *MSNBC.com* (March 23, 2007), accessed November 13, 2011, *http://www.msnbc.msn.com.*

57. Bill Tomson, "Antibiotics in Livestock Feed Raises Concerns," *Wall Street Journal* (May 13, 2011), accessed September 11, 2011, *http://online.wsj.com.*

58. Animal Health Institute, "Animal Antibiotics: Keeping Animals Healthy and Our Food Safe," 4, accessed November 15, 2011, *http://www.ahi.org* (emphasis added).

59. Liggett & Myers Tobacco Company, "Nose, Throat and Accessory Organs Not Adversely Affected by Smoking Chesterfields," advertisement in *Life Magazine* (December 1, 1952), accessed November 15, 2011, *http://www.vintageadsandstuff.com.*

60. R. Smither et al., "Antibiotic Residues in Meat in the United Kingdom: An Assessment of Specific Tests to Detect and Identify Antibiotic Residues," *Journal of Hygiene* 85 (1980): 359–69; Joint Expert Advisory Committee on Antibiotic Resistance, Australia, "The Use of Antibiotics in Food-Producing Animals: Antibiotic Resistance in Animals and Humans" (1999), accessed November 15, 2011, *http://www.health.gov.au*; Frederick W. Oehme, "Significance of Chemical Residues in United States Food-Producing Animals," *Toxicology* 1, no. 3 (1973): 205–15.

61. US Department of Health and Human Services, "NARMS 2009 Executive Report," 52, 66.

62. Ibid., 98.

63. Inge van Loo et al., "Emergence of Methicillin-resistant *Staphylococcus Aureus* of Animal Origin in Humans," *Emerging Infectious Diseases* 13, no. 12 (2007): 1834–39; Andrew E. Waters et al., "Multidrug-Resistant *Staphylococcus Aureus* in U.S. Meat and Poultry," *Clinical Infectious Diseases* 52, no. 10 (2011): 1227–30.

64. Loo et al., "Emergence of Methicillin-resistant *Staphylococcus Aureus.*"

65. Waters et al., "Multidrug-Resistant *Staphylococcus Aureus,*" 1228.

66. S. H. Swan et al., "Semen Quality of Fertile U.S. Males in Relation to Their Mothers' Beef Consumption During Pregnancy," *Human Reproduction* 22, no. 6 (2007): 1497–1502.

67. Lillian Conde de Borrego, "An Epidemic of Precocious Development in Puerto Rican Children," *Journal of Pediatrics* 107, no. 3 (1985): 393–96; Cornell University, "Consumer Concerns about Hormones in Food" (2000), accessed October 8, 2011, *http://envirocancer.cornell.edu.*

68. In inflation-adjusted dollars, the US costs related to *E. coli* and salmonella poisoning are $3.4 billion annually. Salmonella costs are $2.7 billion in 2010 dollars, or $2.9 billion in 2012 dollars; *E. coli* costs are $488 million in 2010

dollars, or $518 million in 2012 dollars. $2.9 billion + $518 million = $3.4 billion. While some poisoning cases are related to pathogens transmitted indirectly by vegetables (although, of course, always originating in animals), in the absence of data showing how much of the total is related to animal foods, it's reasonable to estimate that half of the total, or $1.7 billion, is attributable to pathogens transmitted directly by animal foods. USDA Economic Research Service, "Foodborne Illness Cost Calculator" (2012), accessed October 27, 2012, *http://webarchives.cdlib.org.*

69. US expenses related to antibiotic resistance in humans, including health care costs and lost wages, are estimated at $35 billion in 2000 dollars ($47.2 billion in 2012 dollars). (Rebecca R. Roberts et al., "Hospital and Societal Costs of Antimicrobial-Resistant Infections in a Chicago Teaching Hospital: Implications for Antibiotic Stewardship," *Clinical Infectious Diseases* 49, no. 8 [2009]: 1175–84; PR Newswire, "Antibiotic-Resistant Infections Cost the U.S. Healthcare System in Excess of $20 Billion Annually" [2000], accessed January 11, 2012, *http://www.prnewswire.com.*) But how much of this total is attributable to the use of antibiotics in animals, and how much results from humans' personal use of antibiotics? Eighty percent of the antibiotics used in the United States is fed to, or injected into, farm animals. These animal antibiotics have a significant, but currently unmeasured, effect on human health. The National Academies' Institute of Medicine, for example, has noted that "a decrease in the inappropriate use of antimicrobials in human medicine alone is not enough" to address antimicrobial resistance in humans. "Substantial efforts must be made to decrease inappropriate overuse of antimicrobials in animals and agriculture as well." (Mark S. Smolinski, Margaret A. Hamburg, and Joshua Lederberg, eds., *Microbial Threats to Health: Emergence, Detection, and Response* [Washington, DC: National Academies Press, 2003], 207.) In the absence of relevant calculations in the literature, and in light of the significant portion of antibiotics used on animals and the numerous cited instances of antibiotic-resistant diseases passing from animals to humans, it is reasonable to estimate that half of the total health care costs related to antibiotic resistance is attributable to dosing animals. Accordingly, $23.6 billion ($47.2 billion ÷ 2) is used as the relevant annual figure.

70. Adjusted for inflation, $253 billion in 1980 dollars is $706 billion in 2012 dollars. Henry J. Kaiser Family Foundation, "U.S. Healthcare Costs," accessed January 1, 2012, *http://www.kaiseredu.org.*

71. Adjusted for inflation, $444.2 billion in 2008 dollars is $477 billion in 2012 dollars. Paul A. Heidenreich et al., "Forecasting the Future of Cardiovascular Disease in the United States: A Policy Statement from the American Heart Association," *Circulation* 123 (2011) 933–44.

72. Adjusted for inflation, $226.8 billion in 2007 dollars is $253 billion in 2012 dollars. American Cancer Society, "Cancer Facts & Figures 2012," accessed May 24, 2012, *http://www.cancer.org.*

73. The total 2007 cost associated with diabetes was $174 billion; type 2 diabetes (the type associated with eating animal foods) accounts for 95 percent of the total, or $165.3 billion. After adjustment for inflation, this figure is $184 billion in 2012 dollars. American Diabetes Association, "Economic Costs of Diabetes in the U.S. in 2007," *Diabetes Care* 31, no. 3 (2008); US Centers for Disease Control and Prevention, "Diabetes—Success and Opportunities for Population-Based Prevention and Control: At a Glance 2010," accessed October 18, 2011, *http://www.cdc.gov*.

74. Regarding heart disease, *see, for example,* Iqbal et al., "Dietary Patterns and Acute Myocardial Infarction," abstract: "An unhealthy dietary intake . . . accounts for approximately 30% of the population-attributable risk" of acute myocardial infarction. Regarding cancer, *see, for example,* Walker, "Diet in the Prevention of Cancer," abstract: "Diet is considered responsible for about a third of cases of cancer. . . ." Regarding type 2 diabetes, one study finds nine in ten cases could be avoided by lifestyle changes including eating less saturated fat and red and processed meat. (Dariush Mozaffarian et al., "Lifestyle Risk Factors and New-Onset Diabetes Mellitus in Older Adults," *Archives of Internal Medicine* 169, no. 8 [2009]: 798–807.) This study also found that a healthier diet was associated with a 35 percent lower diabetes risk. *See also* Daniel M. Keller, "High-Protein Diet Raises Type 2 Diabetes Risk," (lecture, European Association for the Study of Diabetes 47th Annual Meeting, Berlin, Germany, 2011) accessed October 20, 2011, *http://www.medscape.com*. Those eating the most animal protein had 37 percent higher diabetes risk than those eating the least.

75. The math is as follows, with heart disease costs first, then cancer, and finally type 2 diabetes: (477 x 0.3) + (253/3) + (184/3) = 288.8.

Chapter 7

1. "Huge Spill of Hog Waste Fuels an Old Debate in North Carolina," *New York Times* (June 25, 1995).

2. Michael Mallin, "Impacts of Industrial Animal Production on Rivers and Estuaries," *American Scientist* 88 (2000): 26–37.

3. Merritt Frey, Rachel Hopper, and Amy Fredregill, "Spills and Kills: Manure Pollution and America's Livestock Feedlot" (Washington, DC: Clean Water Network, 2000).

4. Quoted in "Huge Spill of Hog Waste."

5. Quoted in James McWilliams, *Just Food: Where Locavores Get It Wrong and How We Can Eat Responsibly* (New York: Back Bay Books, 2009), 7.

6. United Nations Environment Program, "Universal Ownership: Why Environmental Externalities Matter to Institutional Investors" (2010), accessed January 17, 2012, *http://www.unpri.org*.

7. Henning Steinfeld et al., *Livestock's Long Shadow: Environmental Issues and Options* (Rome: Food and Agriculture Organization of the United Nations, 2006), xx.

8. Joey Papa, "Styrofoam vs. Paper Cups: Which is More Eco-Friendly?" Recycling: Keep It Out of the Landfill (2010), accessed November 21, 2011, *http://1800recycling.com*.

9. Michael Pollan, *The Omnivore's Dilemma: A Natural History of Four Meals* (New York: Penguin Press, 2006); Joel Salatin, *Pastured Poultry Profits* (Polyface, Inc., 1993).

10. Pollan, *Omnivore's Dilemma*.

11. Comparing the grain inputs fed to chickens to the farm's total meat output for all animals (cattle, chickens, turkeys, pigs, rabbits, and eggs), Merberg finds that caloric input exceeds caloric output by roughly 43 percent. I narrow the comparison to just chicken feed inputs versus chicken-related outputs (eggs, broilers, and stewing hens). Adam Merberg, "The Free Lunch," *Say What, Michael Pollan* (blog) (2011), accessed December 21, 2011, *http://saywhatmichaelpollan.wordpress.com*.

12. David Pimentel and Marcia Pimentel, "Sustainability of Meat-Based and Plant-Based Diets and the Environment," *American Clinical Journal of Nutrition* 78, no. 3 (2003): 6605–35.

13. USDA Economic Research Service, "Loss-Adjusted Food Availability of Meat, Poultry, Fish, Eggs and Nuts" (2012), accessed September 28, 2012, *http://www.ers.usda.gov*.

14. US Census Bureau, "American FactFinder" online database, accessed September 28, 2012, *http://factfinder2.census.gov*.

15. Ruben N. Lubowski et al., "Major Uses of Land in the United States, 2002," USDA Economic Research Service (2006), accessed August 19, 2012, *http://www.ers.usda.gov*.

16. Richard A. Oppenlander, *Comfortably Unaware: Global Depletion and Food Responsibility . . . What You Choose to Eat is Killing Our Planet* (Minneapolis: Langdon Street Press, 2011), Kindle version.

17. Adam Merberg: "The Grain Inputs on Polyface Farm: Joel Salatin's Take," *Say What, Michael Pollan* (blog) (2011), accessed October 27, 2012, *http://saywhatmichaelpollan.wordpress.com*.

18. Organic Trade Association, "Industry Statistics and Projected Growth" (2001), accessed November 21, 2011, *http://www.ota.com*.

19. C. Foster et al., "Environmental Impacts of Food Production and Consumption: A Report to the Department for Environment Food and Rural Affairs," Eldis (2006), accessed August 20, 2011, *http://www.eldis.org*.

20. Data expressed in hectares converted to acres. A. G. Williams, E. Audsley, and D. L. Sandars, "Determining the Environmental Burdens and Resource Use in the Production of Agricultural and Horticultural Commodities"

(2006), Main Report, UK Department of Environment, Food, and Rural Affairs Research Project IS0205.

21. Williams, Audsley, and Sandars, "Environmental Burdens in Production of Agricultural Commodities"; David Pimentel and Marcia Pimentel, *Food, Energy, and Society* (Niwot, CO: Colorado University Press, 1996).

22. US Environmental Protection Agency, "Methane: Science" (2010), accessed October 27, 2012, *http://epa.gov*.

23. L. A. Harper et al., "Direct Measurements of Methane Emissions from Grazing and Feedlot Cattle," *Journal of Animal Science* 77, no. 6 (1999): 1392–1401.

24. ChartsBin, "Total Water Use per Capita by Country," accessed December 23, 2012, *http://chartsbin.com*.

25. Assuming the animal weighs 1,200 pounds; metric units converted to imperial. T. Oki et al., "Virtual Water Trade to Japan and in the World" (presentation, International Expert Meeting on Virtual Water Trade, Netherlands, 2003), accessed November 22, 2011, *http://www.waterfootprint.org*.

26. Pimentel and Pimentel, *Food, Energy and Society*.

27. Peter H. Gleick et al., *The World's Water: The Biennial Report on Freshwater Resources* vol. 7 (Washington, DC: Island Press, 2011), 337–38; USDA National Nutrient Database for Standard Reference, "Release 25: Content of Selected Foods per Protein (g) Common Measure, Sorted Alphabetically" (2012), accessed December 23, 2012, *https://www.ars.usda.gov*.

28. World Health Organization, *Safer Water, Better Health: Costs, Benefits, and Sustainability of Interventions to Protect and Promote Health* (2008), accessed December 19, 2011, *http://www.who.int*.

29. National Oceanic and Atmospheric Administration, "North American Drought Monitor—July 2012" (August 2012), accessed August 27, 2012, *http://www1.ncdc.noaa.gov*.

30. Ibid., 4.

31. As measured by the Palmer Hydrological Drought Index. National Oceanic and Atmospheric Administration, "National Drought Overview," State of the Climate: Drought (November 2011), accessed December 19, 2011, *http://www.ncdc.noaa.gov*.

32. Pimentel and Pimentel, "Meat-based and Plant-based Diets."

33. Christopher L. Weber and H. Scott Matthews, "Food-Miles and the Relative Climate Impacts of Food Choices in the United States," *Environmental Science and Technology* 42, no. 10 (2008): 3508–13.

34. Rich Pirog et al., "Food, Fuel, and Freeways: An Iowa Perspective on How Far Food Travels, Fuel Usage, and Greenhouse Gas Emissions," Leopold Center for Sustainable Agriculture (2001), accessed December 21, 2011, *http://www. leopold.iastate.edu*.

35. Caroline Saunders and Andrew Barber, "Carbon Footprints, Life Cycle Analysis, Food Miles: Global Trade Trends and Market Issues," *Political Science* 60, no. 1 (2008): 73–88.

36. Ibid.

37. Ibid.

38. Ibid., 87.

39. McWilliams, *Just Food*, 214.

40. Peter Miller and William E. Rees, "Introduction," in *Ecological Integrity: Integrating Environment, Conservation and Health*, eds. David Pimentel, Laura Westra, and Reed F. Noss (Washington, DC: Island Press, 2000), 6.

41. Dennis vanEngelsdorp et al., "A Survey of Honey Bee Colony Losses in the U.S., Fall 2007 to Spring 2008," *PLoS ONE* 3, no. 12 (2008): e4071, accessed October 24, 2011, *http://www.plosone.org*.

42. Ibid.

43. George Raine, "Many Causes Blamed for Honeybee Die-off: Colony Collapse Disorder Could Cost $15 Billion," *San Francisco Chronicle* (June 1, 2007).

44. US Department of Agriculture, *The Second RCA Appraisal: Soil, Water, and Related Resources on Nonfederal Land in the United States: Analysis of Conditions and Trends* (Washington, DC: USDA, 1989).

45. D. Pimentel et al., "Environmental and Economic Costs of Soil Erosion and Conservation Benefits," *Science* 267, no. 5201 (1995): 1117–23.

46. Steinfeld et al., *Livestock's Long Shadow*, 168.

47. Research found $17 billion in "off-site" losses in 1992 dollars, or $27.9 billion in 2012 dollars. I ignore "on-site" losses of roughly $27 billion—$44.2 billion in current dollars—because these costs are internalized and incurred by the producers. 27.9 x 0.55 = 15.4. Pimentel et al., "Environmental and Economic Costs of Soil Erosion," 1120.

48. Robert Goodland and Jeff Anhang, "Livestock and Climate Change: What if the Key Actors in Climate Change Are . . . Cows, Pigs and Chickens?" World Watch (November/December 2009), accessed October 25, 2011, *http://www. worldwatch.org*.

49. Noreen Malone, "This Weather Is What Global Warming Looks Like," *New York Magazine* (July 3, 2012), accessed September 9, 2012, *http://nymag.com*.

50. This report, commonly known as AR4, proposes several scenarios for mitigating CO_2 emissions by 2030, at various levels of aggressiveness and related cost. The most aggressive scenario, and arguably the most sensible, calls for stabilizing CO_2 emissions at 445–535 parts per million (ppm) by 2030. Stabilization at the midpoint of this range, 490 ppm, would mean the average planetary temperature would rise above preindustrial levels by 4.7°F and above today's levels by 3.3°F. Although it's almost 100 points higher than today's level of 395 ppm, acceleration in the ppm growth rate means we'll get there much faster than we think. Incidentally, the present concentration of CO_2 in the atmosphere is the highest it's been in the past 650,000 years. The rise in CO_2 over the past century increased the planet's average temperature by 1.3°F. Not only does CO_2's ppm increase each year, but the rate of increase also increases steadily. Thus, before 2000, CO_2 increased by an average of 1.5

ppm per year; from 2000 to 2007 it increased at 2.1 ppm per year; and from 2009 to 2010 it increased by 3.0 ppm. This trend in rate increases suggests we'll soon see annual ppm increases of 5.0 and higher. Intergovernmental Panel on Climate Change, "Climate Change 2007: Synthesis Report" (2007), accessed October 27, 2012, *http://www.ipcc.ch.*

51. $15.1 trillion x 0.0012 x 0.51 = $9.2 billion in 2011 dollars or $9.4 billion in current dollars.

52. Pesticide costs: Research finds $9.6 billion (in 2003 dollars) in costs associated with pesticide use. (D. Pimentel, "Environmental and Economic Costs of the Application of Pesticides Primarily in the United States," *Environment, Development and Sustainability* 7 [2005]: 229–52.) The USDA estimates that 52 percent of US cropland is dedicated to raising feed crops; 52 percent of the total pesticide costs attributable to feed crops is $5 billion, or an inflation-adjusted $6.3 billion. (US Department of Agriculture, "Major Uses of Land in the United States, 2002/EIB-14," USDA Economic Research Service [2002], 20, accessed September 28, 2012, *http://www.ers.usda.gov.*)

    Fertilizer costs: Research finds total costs of $2.2 billion arising from all US agriculture in 2008 dollars; the 52 percent of this total attributable to feed crops is $1.1 billion, or $1.2 billion in current dollars. Adding this figure to the $6.3 billion in pesticide costs yields $7.5 billion. (Walter K. Dodds et al., "Eutrophication of U.S. Freshwaters: Analysis of Potential Economic Damages," *Environmental Science and Technology* 43, no. 1 [2009].)

53. "New Fear from Hog Lots: Odor May Spread Illness—Evidence Mounts that Neighbors Are at Risk," *The Des Moines Register* (October 25, 1998).

54. Karen McDonald, "Hog Farm Smell Has Residents Racing for Windows," *Peoria Journal Star* (September 11, 2005).

55. Joseph A. Herriges, Silvia Secchi, and Bruce A. Babcock, "Living with Hogs: The Impact of Livestock Facilities on Rural Residential Property Values" (working paper, 03-WP 342, Center for Agricultural and Rural Development, Iowa State University 2003), accessed October 25, 2011, *http://www. ncifap.org*; Raymond B. Palmquist, Fritz M. Roka, and Tomislav Vukina, "Hog Operations, Environmental Effects, and Residential Property Values," *Land Economics* 73, no. 1 (1997): 114–24.

56. Research found that the average decline in real property values within a three-mile radius from each CAFO in Missouri was $2.68 million in 1999 dollars, or $3.7 million today. (Mubarak Hamed, Thomas G. Johnson, and Kathleen K. Miller, *The Impacts of Animal Feeding Operations on Rural Land Values* [Report R-99-02, College of Agriculture, University of Missouri-Columbia, 1999].) Extrapolating this average to the 18,800 CAFOs the EPA says operate in the United States yields $69.6 billion. (US EPA, "Fact Sheet: Concentrated Animal Feeding Operations Proposed Rulemaking June 2006" [2006], accessed July 5, 2012, *http://www.epa.gov.*) Amortizing this cost figure over the 27.5-year useful life of residential property (as decreed by the IRS) yields an annualized, externalized cost of $2.5 billion in current dollars.

57. The EPA says animal feed operations produce more than 500 million tons of manure (1 trillion pounds) yearly. US Environmental Protection Agency, "National Pollutant Discharge Elimination System Permit Regulation and Effluent Limitation Guidelines and Standards for Concentrated Animal Feeding Operations (CAFOs)" 68 FR 7176-01 (2003).

58. US Environmental Protection Agency, "National Water Quality Inventory 2000 Report" (2000), accessed October 24, 2011, *http://www.epa.gov.*

59. Researchers have found that the cost to spread waste over cropland instead of storing it would be $1.2 billion per year in 2005 dollars—$1.4 billion in 2012 dollars. (Marcel Aillery et al., "Managing Manure to Improve Air and Water Quality," *Economic Research Report* 9, US Department of Agriculture, Economic Research Service [2005]: 26, accessed October 24, 2011, *http://www.ers.usda.gov.*) Further, research estimates that the cost to repair leaky lagoons is $4.1 billion in 2003 dollars ($5.1 billion in 2012 dollars). (C. Volland, J. Zupancic, and J. Chappell, "Cost of Remediation of Nitrogen-Contaminated Soils under CAFO Impoundments," *Journal of Hazardous Substance Research* 4, no. 3 [2003]: 1–18; Doug Gurian-Sherman, "CAFOs Uncovered: The Untold Costs of Confined Animal Feeding Operations" [2008], 6, accessed October 27, 2012, *http://www.ucsusa.org.*) Because even manure destined to be spread on cropland must be stored for some time to allow solids to decompose, these two costs—spreading and repairing—are not mutually exclusive. To convert the repair cost to an annual figure, it is amortized over the repair's useful life of five years (per the IRS), and assuming that normal wear and tear requires further repairs, it is expected that similar costs will arise every five years. This yields an annual repair cost of about $1 billion, and added to the spreading costs, a total annual cost to deal with manure problems of about $2.4 billion.

60. US Government Accountability Office, "Concentrated Animal Feeding Operations: EPA Needs More Information and a Clearly Defined Strategy to Protect Air and Water Quality from Pollutants of Concern" GAO-08-944 (2008), 6.

61. William W. Simpkins et al., "Potential Impact of Earthen Waste Storage Structures on Water Resources in Iowa," *Journal of the American Water Resources Association* 38, no. 3 (2002): 759–71.

62. US Geological Survey, "Fact Sheet FS-027-02, Pharmaceuticals, Hormones, and Other Organic Wastewater Contaminants in U.S. Streams" (2002), accessed October 24, 2011, *http://toxics.usgs.gov.*

63. US Government Accountability Office, "Livestock Agriculture: Increased EPA Oversight Will Improve Environmental Program for Concentrated Animal Feeding Operations" (report to the Ranking Member, Committee on Agriculture, Nutrition and Forestry, US Senate, 2003), 3.

64. Ibid.

65. Edgar G. Hertwich et al., "Assessing the Environmental Impacts of Consumption and Production: Priority Products and Materials" (report, Working Group on the Environmental Impacts of Products and Materials to the International Panel for Sustainable Resource Management, United Nations Environment Program, 2010), 80, accessed December 20, 2011, *www.unep.fr.*

66. US Environmental Protection Agency, "Major Crops Grown in the United States," accessed December 20, 2011, *http://www.epa.gov*; Lester Brown, "Soybeans Threaten Amazon Rainforest," Grist (2010), accessed October 27, 2012, *http://grist.org.*

67. McWilliams, *Just Food*, 119.

68. Christian J. Peters, Jennifer L. Wilkins, and Gary W. Fick, "Testing a Complete Diet Model for Estimating the Land Resource Requirements of Food Consumption and Agricultural Carrying Capacity: The New York State Example," *Renewable Agriculture and Food Systems* 22, no. 2 (2007): 145–53.

69. Square kilometers converted to acres. Brazil Ministry of Science and Technology, National Institute for Space Research, "INPE Estimates a Reduction of 11% in Amazon Deforestation" (2011), accessed December 20, 2011, http://*www.inpe.br.*

70. P. M. Fearnside, "Deforestation in Brazilian Amazonia: History, Rates and Consequences," *Conservation Biology* 19, no. 3 (2005): 680–88.

71. Kilograms converted to pounds. Pimentel and Pimentel, "Meat-Based and Plant-Based Diets."

72. Ibid.; UN Food and Agriculture Organization, "The State of Food Insecurity in the World: Addressing Food Insecurity in Protracted Crises" (2010), accessed December 21, 2011, *http://www.fao.org.*

73. Vaclav Smil, *Feeding the World: A Challenge for the Twenty-First Century* (Cambridge, MA: MIT Press, 2001).

74. US Environmental Protection Agency, "Ag 101: Land Use Overview," National Agriculture Center, accessed December 21, 2011, *http://www.epa. gov.*

75. McWilliams, *Just Food*, 120.

76. Hertwich et al., "Assessing the Environmental Impacts of Consumption and Production."

77. Ibid., 82.

78. US Environmental Protection Agency, "Methane: Sources and Emissions" (2011), accessed October 29, 2012, *http://www.epa.gov.*

Chapter 8

1. James D. Rose, "Do Fish Feel Pain?" (2000), accessed July 19, 2012, *http://www.coloradotu.org.*

2. Matthew Scully, *Dominion: The Power of Man, the Suffering of Animals, and the Call to Mercy* (New York: St. Martin's Griffin, 2002), 258–59.

3. *See, for example*, F. Bailey Norwood and Jayson L. Lusk, *Compassion by the Pound: The Economics of Farm Animal Welfare* (New York: Oxford University Press, 2011), 344–45: "75 percent [of survey respondents] say they would vote for a law in their state requiring farmers to treat their animals better."

4. Ron Torell, "Cow Camp Chatter: Replacement Rate," *Sage Signals: Voice of the Nevada Livestock Industry* (November, 2011): 6, accessed August 22, 2012, *http://www.nevadacattlemen.org*.

5. Nicholas Fontaine, *Memories Pour Server a l'Histoire de Port-Royal* (Cologne, 1738): 52–53; quoted in *From Beast-Machine to Man-Machine: The Theme of Animal Soul in French Letters from Descartes to La Mettrie*, L. Rosenfield (New York: Oxford University Press, 1940).

6. Lynne Sneddon, "The Evidence for Pain in Fish: The Use of Morphine as an Analgesic," *Applied Animal Behaviour Science* 83, no. 2 (2003): 153–62.

7. D. A. Denton et al., "The Role of Primordial Emotions in the Evolutionary Origin of Consciousness," *Consciousness and Cognition* 18, no. 2 (2009): 500–14.

8. Lynne U. Sneddon, Victoria A. Braithwaite, and Michael J. Gentle, "Do Fishes Have Nociceptors? Evidence for the Evolution of a Vertebrate Sensory System," *Proceedings of the Royal Society B: Biological Sciences* 270, no. 1520 (2003): 1115–21.

9. Sneddon, "Pain in Fish."

10. Ibid.

11. Janicke Nordgreen et al., "Thermonociception in Fish: Effects of Two Different Doses of Morphine on Thermal Threshold and Post-Test Behaviour in Goldfish (*Carassius Auratus*)," *Applied Animal Behavior Science* 119, no. 1 (2009): 101–7.

12. Quoted in Brian Wallheimer, "Fish May Actually Feel Pain and React to It Much Like Humans," *Purdue University News* (April 29, 2009).

13. T. G. Pottinger and A. D. Pickering, "Genetic Basis to the Stress Response: Selective Breeding for Stress Tolerant Fish," in *Fish Stress and Health in Aquaculture*, ed. G. K. Iwama et al. (Cambridge, UK: Cambridge University Press, 1997).

14. Aquaculture Innovation Network, "Cataracts in Farmed Fish," accessed August 26, 2011, *http://www.aquamedia.org*.

15. P. J. Midtlyng et al., "Current Research on Cataracts in Fish," *Bulletin of the European Association of Fish Pathologists* 19, no. 6 (1999): 299–301.

16. Mark D. Powell and Steve F. Perry, "Respiratory and Acid-Base Pathophysiology of Hydrogen Peroxide in Rainbow Trout (*Oncorhynchus mykiss Walbaum*)," *Aquatic Toxicology* 37, no. 2 (1997): 99–112.

17. N. W. Ross et al., "Changes in Hydrolytic Enzyme Activities of Naive Atlantic Salmon (*Salmo salar*) Skin Mucus Due to Infection with the Salmon Louse

(*Lepeophtheirus salmonis*) and Cortisol Implantation," *Diseases of Aquatic Organisms* 41, no. 1 (2000): 43–51; S. C. Johnson and L. J. Albright, "Effects of Cortisol Implants on the Susceptibility and the Histopathology of the Responses of Naive Coho Salmon Oncorhynchus kisutch to Experimental Infection with *Lepeophtheirus salmonis* (Copepoda: Caligidae)," *Diseases of Aquatic Organisms* 14 (1992): 195–205.

18. D. Mackay, "Perspectives on the Environmental Effects of Aquaculture," (presentation, Aquaculture Europe Conference, Norway, August 1999).

19. S. C. Kestin, S. B. Wotton, and S. Adams, "The Effect of CO_2, Concussion or Electrical Stunning of Rainbow Trout (*Oncorhynchus mykiss*) on Fish Welfare," in *Quality in Aquaculture, Special Publication* 23, eds. N. Svennevig and A. Krogdahl (Gent, Belgium: European Aquaculture Society, 1995); D. H. F. Robb et al., "Commercial Slaughter Methods Used on Atlantic Salmon: Determination of the Onset of Brain Failure by Electroencephalography," *Veterinary Record* 147, no. 11 (2000): 298–303.

20. European Food Safety Authority, "Scientific Report of the Scientific Panel on Animal Health and Welfare" (2004), accessed August 27, 2011, *http://www. efsa.europa.eu.*

21. S. C. Kestin, D. H. F. Robb, and J. W. van de Vis, "Protocol for Assessing Brain Function in Fish and the Effectiveness of Methods Used to Stun and Kill Them," *Veterinary Record* 150 (2002): 302–7.

22. Robb et al., "Commercial Slaughter Methods Used on Atlantic Salmon"; E. Lambooij, R. J. Kloosterboer, M. A. Gerritzen, and J. W. van de Vis, "Head-Only Electrical Stunning and Bleeding of African Catfish (*Clarias gariepus*): Assessment of Loss of Consciousness," *Animal Welfare* 13, no. 1 (2004): 71–76.

23. S. C. Kestin, S. B. Wotton, and N. G. Gregory, "Effect of Slaughter by Removal from Water on Visual Evoked Activity in the Brain and Reflex Movement of Rainbow Trout (*Oncorhynchus mykiss*)," *Veterinary Record* 128, no. 19 (1991): 443–46.

24. Per Olav Skjervold et al., "Live-Chilling and Crowding Stress Before Slaughter of Atlantic Salmon (*Salmo salar*)," *Aquaculture* 192, no. 2 (2001): 265–80.

25. Norwood and Lusk, *Compassion by the Pound*, 298.

26. Unrounded, the figures are $341.53 and $345.09. Ibid.

27. $343.31 x 5 = $1,716.55 in 2011 dollars; the inflation-adjusted total is $1,751.10.

28. Assuming 236.8 million US adults: ($1,751.10 x 236,800,000)/20 = $20,733,024,000.

29. L. R. Mathews and J. Ladewig, "Environmental Requirements of Pigs Measured by Behavioral Demand Functions," *Animal Behavior* 47, no. 3 (1994): 713–19.

Chapter 9

1. Food and Agriculture Organization of the United Nations, "Guidelines to Reduce Sea Turtle Mortality in Fishing Operations" (2010), accessed October 1, 2012, *http://www.fao.org.*

2. *Ctr. for Marine Conservation v. Brown*, 917 F. Supp. 1128, 1136 (S.D. Tex. 1996)

3. Maya Rodriguez, "Federal Regulators, Shrimpers at Odds over Use of Turtle Excluder Devices," *WWWLTV.com Eyewitness News* (June 5, 2012), accessed June 15, 2012, *http://www.wwltv.com.*

4. Food and Agriculture Organization of the United Nations, "Guidelines to Reduce Sea Turtle Mortality."

5. FAO states that 23,436 fishing ships have IHS-F (IMO) numbers, a numbering status reserved for ships of 100 tons or more. Food and Agriculture Organization of the United Nations, "The State of World Fisheries and Aquaculture" (2010), 107, accessed October 1, 2012, *http://www.fao.org.*

6. Canada, Russia, Norway, Iceland, Greenland, and New Zealand all ban discards to some extent, and the European Union recently adopted a limited ban with effect in 2014. Discard bans require that bycatch be landed and counted against quotas, which provide incentives to fish with gear and in areas that minimize bycatch. While bans are clearly better than ignoring the issue, their effectiveness is often limited by cheating, continuing discards of noncontrolled species, and questions as to how to handle landed bycatch. (Ivor Clucas, "A Study of the Options for Utilization of Bycatch and Discards from Marine Capture Fisheries," *UN FAO Fisheries Circular* 928 FIIU/C928 [1997], accessed June 23, 2012, *http://www.fao.org.*) Nevertheless, the Norwegian government believes its ban is working. "The very existence of the rule," says the Norwegian Ministry of Fisheries and Coastal Affairs, "has proved beneficial in changing fishermen's attitudes and discouraging the practice of discarding." (Norwegian Ministry of Fisheries and Coastal Affairs, "Norwegian Fisheries Management, Our Approach on Discard of Fish," accessed June 23, 2012, *www.regjeringen.no.*)

7. International Union for Conservation of Nature, "IUCN Red List of Threatened Species," accessed August 25, 2011, *http://www.iucnredlist.org.*

8. Nigel Brothers, "Albatross Mortality and Associated Bait Loss in the Japanese Longline Fishery in the Southern Ocean," *Biological Conservation* 55, no. 3 (1991).

9. R. W. D. Davies et al., "Defining and Estimating Global Marine Fisheries Bycatch," *Marine Policy* 33, no. 4 (2009): 661–72.

10. Annual worldwide catch is estimated at 86 million tons, or 172 billion pounds. (Michael Parfit, "Diminishing Returns," *National Geographic* [November 1995].) Davies et al. estimate that 40.4 percent of that total, or 69.5 billion pounds, is discarded each year as bycatch. Dividing this figure by

365 reveals that roughly 190.4 million pounds of bycatch is discarded each day. (Davies et al., "Estimating Bycatch.")

11. Dayton L. Alverson et al., *A Global Assessment of Fisheries Bycatch and Discards* (Rome: Food and Agriculture Organization of the United Nations, 1994).
12. Ibid.
13. Ibid.
14. Ibid.
15. Ibid.
16. Ibid (emphasis added).
17. World Bank and United Nations Food and Agriculture Organization, "The Sunken Billions: The Economic Justification for Fisheries Reform" (2009), 21, accessed October 1, 2012, *http://www.worldbank.org.*
18. US fishing subsidies total $1.8 billion in 2003 dollars, or $2.3 billion today. Sumaila et al., "Re-Estimation of Fisheries Subsidies."
19. According to NOAA, US seafood landings in 2010 totaled 8.2 billion pounds and were worth $4.5 billion (or $4.8 billion in 2012 dollars). $4.8 billion / 8.2 billion pounds = $0.59 per pound. The $2.3 billion annual subsidy is 0.48 of the value of total landings of $4.8 billion, and as $0.59 x 0.48 = 0.28 (rounded), this implies a subsidy value of $0.28 for each pound landed. (Note that because of value added in fish processing, the price of fish at their first introduction to human markets, as landings, is much lower than when later offered as a retail product.) National Oceanic and Atmospheric Administration, "U.S. Domestic Seafood Landings and Values Increase in 2010" (2011), accessed October 30, 2012, *http://www.noaanews.noaa.gov.*
20. Sumaila et al., "Re-Estimation of Fisheries Subsidies."
21. World Bank and UN Food and Agriculture Organization, *The Sunken Billions*, 31.
22. Boris Worm et al., "Impacts of Biodiversity Loss."
23. Ibid.
24. Ibid.
25. United Nations Environment Program Millennium Ecosystem Assessment, "Collapse of Atlantic Cod Stocks Off the East Coast of Newfoundland in 1992" (2005), accessed October 1, 2012, *http://www.grida.no.*
26. Lan T. Gien, "Land and Sea Connection: The East Coast Fishery Closure, Unemployment and Health," *Canadian Journal of Public Health* 91, no. 2 (2000): 121–24.
27. Erlingur Hauksson and Valur Bogason, "Comparative Feeding of Grey (*Halichoerus grypus*) and Common Seals (*Phoca vitulina*) in Coastal Waters of Iceland, with a Note on the Diet of Hooded (*Cystophora cristata*) and Harp Seals (*Phoca groenlandica*)," *Journal of Northwest Atlantic Fishery Science* 22 (1997).

28. National Oceanic and Atmospheric Administration, "Fishwatch: U.S. Seafood Facts," accessed October 1, 2012, *http://www.fishwatch.gov*.
29. Matthias Halwart, Doris Soto, and J. Richard Arthur, eds., "Cage Aquaculture: Regional Reviews and Global Overview" (technical paper no. 498, UN FAO Fisheries, Rome, 2007).
30. Philip Lymberly, "In Too Deep: The Welfare of Intensively Farmed Fish," Compassion in World Farming Trust (2002), accessed October 1, 2012, at *http://www.ciwf.org.uk*.
31. C. Sommerville, "Parasites of Farmed Fish," in *Biology of Farmed Fish*, eds. K. D. Black and A. D. Pickering (Sheffield, UK: Sheffield Academic Press, 1998).
32. A. Mustafa, G.A. Conboy, and J. F. Burka, "Life-Span and Reproductive Capacity of Sea Lice, *Lepeophtheirus salmonis*, under Laboratory Conditions," *Aquaculture Association of Canada Special Publication* 4 (2000): 113–14.
33. M. Krkošek et al., "Epizootics of Wild Fish Induced by Farm Fish," *Proceedings of the National Academy of Sciences* 103, no. 42 (2006): 15506–10.
34. M. Krkošek et al., "Declining Wild Salmon Populations in Relation to Parasites from Farm Salmon," *Science* 318, no. 5857 (2007): 1772–75.
35. Cornelia Dean, "Saving Wild Salmon, in Hopes of Saving the Orca," *New York Times* (November 4, 2008).
36. R. J. Goldburg, M. S. Elliott, and R. L. Naylor, *Marine Aquaculture in the United States: Environmental Impacts and Policy Options* (Arlington, VA: Pew Oceans Commission, 2001).
37. W. Ernst et al., "Dispersion and Toxicity to Non-Target Aquatic Organisms of Pesticides Used to Treat Sea Lice on Salmon in Net Pen Enclosures," *Marine Pollution Bulletin* 42, no. 6 (2001): 433–44.
38. M. MacGarvin, "Scotland's Secret: Aquaculture, Nutrient Pollution, Eutrophication and Toxic Blooms," *WWF Scotland* (2000), accessed October 1, 2012, *http://assets.wwf.org.uk*.
39. Quirin Schiermeier, "Fish Farms' Threat to Salmon Stocks Exposed," *Nature* 425, no. 6960 (2003): 753.
40. L. P. Hansen, J. A. Jacobsen, and R. A. Lund, "The Incidence of Escaped Farmed Atlantic Salmon, *Salmo salar* L., in the Faroese Fishery and Estimates of Catches of Wild Salmon," *ICES Journal of Marine Science* 56 (1999): 200–6.
41. Schiermeier, "Fish Farms' Threat Exposed."
42. Rosamond L. Naylor et al., "Effect of Aquaculture on World Fish Supplies," *Nature* 405 (2000): 1017–24.
43. Margot L. Stiles et al., "Hungry Oceans: What Happens When the Prey is Gone?" Oceana (2009), 13, accessed June 22, 2012, *http://oceana.org*.
44. Ibid., 18.
45. Ibid.

46. Quoted in Dan Shapley, "3 Major Reports Paint Same Picture: Ocean Fish Are Rapidly in Decline," *The Daily Green* (March 4, 2009), accessed June 22, 2012, *http://www.thedailygreen.com*.

47. Elisabeth Rosenthal, "Another Side of Tilapia, the Perfect Factory Fish," *New York Times* (May 2, 2011).

48. Kelly L. Weaver et al., "The Content of Favorable and Unfavorable Polyunsaturated Fatty Acids Found in Commonly Eaten Fish," *Journal of the American Dietetic Association* 108 (2008): 1178–85.

49. Rosenthal, "Another Side of Tilapia."

50. G. Knapp, "The World Salmon Farming Industry," in *The Great Salmon Run: Competition between Wild and Farmed Salmon*, eds. G. Knapp, C. Roheim, and J. Anderson (Washington, DC: World Wildlife Fund, 2007).

51. Rebecca Clausen and Stefano D. Longo, "The Tragedy of the Commodity and the Farce of AquAdvantage Salmon," *Development and Change* 43, no. 1 (2012): 229–51.

52. Rivera-Ferre, "A Chicken and Egg Paradigm?"

53. *See, for example*, Carl Folke et al., "The Ecological Footprint Concept for Sustainable Seafood Production: A Review," *Ecological Applications* 8, no. 1, supplement (1998): S63–S71.

54. Carl Folke, Nils Kautsky, and Max Troell, "The Costs of Eutrophication from Salmon Farming: Implications for Policy," *Journal of Environmental Management* 40 (1994): 173–82.

55. Li Lai and Xian-jin Huang, "Environmental Cost Accounting of Pen Fish Farming in East Tai Lake," *Resources Science* 30, no. 10 (2008): 1579–84, abstract.

56. Rosenthal, "Another Side of Tilapia."

57. NOAA Fisheries Service, "Aquaculture in the United States," accessed June 18, 2012, *http://www.nmfs.noaa.gov*.

58. World Bank and UN FAO, *The Sunken Billions*, 1 (emphasis added).

59. Folke et al., "Costs of Eutrophication"; Lai and Huang, "Cost Accounting of Fish Farming."

60. The 2009 US total production value for fish farming was $1.17 billion, or an inflation-adjusted $1.3 billion. Rick Martin, "Review & Forecast: Fisheries and Aquaculture: Landings Up, but Farmed Production, Revenues Down," *Sea Technology Magazine* (January 2012), accessed June 13, 2012, *http://www.sea-technology.com*.

61. Srinivasan et al. show that that lost landings from overfishing are 23 percent of the combined total of actual and lost landings. As 0.23/(1 - 0.23) = .3 (rounded), this lost-catch wedge is equal to roughly 30 percent of actual landings. U. Thara Srinivasan et al., "Food Security Implications of Global Marine Catch Losses Due to Overfishing," *Journal of Bioeconomics* 12, no. 3 (2010): 183–200.

62. US landings were $4.5 billion in 2010 dollars, or an inflation-adjusted $4.8 billion; hence the lost catch is worth 0.3 x $4.8 = $1.4 billion. (Landings figure: NOAA, "U.S. Domestic Seafood Landings.")
63. NOAA, "US Domestic Seafood Landings."
64. $4.8 billion in total commercial landings plus $1.3 billion in aquaculture output.
65. Robert Costanza et al., "The Value of the World's Ecosystem Services and Natural Capital," *Nature* 387 (1997): 253–60.
66. Worm et al., "Impacts of Biodiversity Loss on Ocean Ecosystem Services."

Chapter 10

1. US farm subsidies, for example, impose huge costs on the developing world, which are beyond this book's scope. But in an article titled "How Much Does It Hurt?" researchers from the International Food Policy and Research Institute show that developing countries lose $7 billion yearly because of US subsidy policy. (Xinshen Diao, Eugenio Diaz-Bonilla, and Sherman Robinson, "How Much Does It Hurt?: The Impact of Agricultural Trade Policies on Developing Countries," International Food Policy Research Institute [2003], accessed July 24, 2012, *http://www.ifpri.org*.) Americans' fish consumption is another activity that offloads heavy costs onto the rest of the world. Because we import the majority of the fish we eat, the estimate of $4.5 billion in externalized costs from fishing doesn't count non-US costs and thus is likely too low by a factor of three or more.

    Then there are the costs that haven't been accurately measured yet but will be someday. These include health care costs related to the many other diseases known to be caused at least in part by eating animal foods, such as Parkinson's, Alzheimer's, Crohn's, arthritis, and others. And environmental costs associated with ecosystem damage we're just beginning to understand and measure, such as Colony Collapse Disorder and disruption of marine ecosystems—which as we've seen, one estimate pegs at $21 trillion.

    Finally, there are the hidden costs that don't count in conventional economics, like cruelty costs incurred by animals themselves. Future alternative approaches to economics may measure the costs to farm animals, as economic actors, of intensive confinement and other industrial practices. And future research will likely provide reason to dramatically increase the estimated health and environmental costs of animal food production. For now, the existence of these numerous additional, uncounted costs is sufficient to show that if anything, the estimate of $414 billion is conservative.
2. Body mass index (BMI) is calculated by dividing weight in kilograms by height in meters squared, or kg/m². S. Tonstad et al., "Type of Vegetarian Diet, Body Weight, and Prevalence of Type 2 Diabetes," *Diabetes Care* 32, no. 5 (2009): 791–96.
3. Ibid., abstract.

4. Jack Norris and Ginny Messina, "Disease Markers of Vegetarians" (2009), table 1, accessed August 19, 2012, *http://www.veganhealth.org.*

5. W. P. Castelli, "Epidemiology of Coronary Heart Disease: the Framingham Study," *American Journal of Medicine* 76, no. 2A (1984): 4–12.

6. Kari Hamerschlag "Meat Eaters Guide to Climate Change + Health," Environmental Working Group (2011), 12, accessed July 29, 2012, *http://www.ewg.org.*

7. Allison Righter, "Ditching Meat One Day a Week: What, Exactly Is the Reduced Risk of Mortality?" (2012), accessed July 29, 2012, *http://www.livablefutureblog.com.*

8. US Centers for Disease Control and Prevention, *Sustaining State Programs for Tobacco Control: Data Highlights 2006* (Atlanta: US Department of Health and Human Services, 2006), accessed October 26, 2011, *http://www.cdc.gov.*

9. F. J. Chaloupka et al., "Tax, Price and Cigarette Smoking: Evidence from the Tobacco Documents and Implications for Tobacco Company Marketing Strategies," *Tobacco Control* 11 (2002): i62–i72, accessed October 26, 2011, *http://tobaccocontrol.bmj.com.*

10. Ann Boonn, "Raising Cigarette Taxes Always Increases State Revenues (and Always Reduces Smoking)" Campaign for Tobacco-Free Kids (2011), accessed August 6, 2012, *http://www.tobaccofreekids.org.*

11. Campaign for Tobacco-Free Kids, "Raising Cigarette Taxes Reduces Smoking, Especially Among Kids (and the Cigarette Companies Know It)" (2007), accessed August 26, 2012, *http://www.tobaccofreekids.org.*

12. US Centers for Disease Control and Prevention, "Consumption Data: Total and Per Capita Adult Yearly Consumption of Manufactured Cigarettes and Percentage Changes in Per Capita Consumption—United States, 1900–2006" (2007), accessed August 26, 2012, *http://www.cdc.gov*; National Cancer Institute, "A Snapshot of Lung Cancer: Incidence and Mortality Rate Trends" (2011), accessed August 26, 2012, *http://www.cancer.gov.*

13. Tax Policy Center, "State and Local Tobacco Tax Revenue, Selected Years 1977–2009" (2011), accessed August 26, 2012, *http://www.taxpolicycenter.org*; Orzechowski and Walker, "The Tax Burden on Tobacco" (2009), accessed August 26, 2012, *http://www.tobaccoissues.com.*

14. Hugh Waters et al., *The Economics of Tobacco and Tobacco Taxation in Mexico* (Paris: International Union against Tuberculosis and Lung Disease, 2010), accessed August 31, 2012, *http://global.tobaccofreekids.org.*

15. Audrey Jacquet, "French Sales of Tobacco Go Up in Smoke as Tax Rises Force Cigarette Buyers Abroad," *The Independent* (July 29, 2004), accessed September 1, 2012, *http://www.independent.co.uk.*

16. Yoree Koh, "Japanese Smokers: Going the Way of the Dodo?" *Wall Street Journal Japan* (blog), October 13, 2011, *http://blogs.wsj.com.*

17. *See* Appendix C, table C1.

18. That is, 0.03 x 0.65 = 0.02 (rounded). The $7.4 billion in revenue is based on the retail sales figure of $251 billion reduced by 2 percent to reflect lower demand, then multiplied by the tax rate of 3 percent. Thus, .03(251 - 0.02(251)) = 7.4.

19. Ann Boonn, "State Cigarette Excise Tax Rates and Rankings" (2011), Campaign for Tobacco-Free Kids, accessed August 6, 2012, *http://www.tobaccofreekids.org.*

20. I'm not the first to propose such a tax. Philosopher and ethicist Peter Singer, for example, proposed a 50 percent tax on meat in a 2009 newspaper editorial. Peter Singer, "Make Meat-Eaters Pay: Ethicist Proposed Radical Tax, Says They're Killing Themselves and the Planet," *New York Daily News* (October 25, 2009).

21. *See* Appendix C, table C7.

22. *See* Appendix C, table C7.

23. *See* Appendix C, table C9.

24. Rob Bluey, "Chart of the Week: Nearly Half of All Americans Don't Pay Income Taxes," *Heritage Network: The Foundry* (blog), February 19, 2012, *http://blog.heritage.org.*

25. *See* chapter 1, table 1.1.

26. Spending in 2013 was $209 billion, higher than the $155 billion budget because some revenue comes from sources other than the US Treasury. US Department of Agriculture, "FY 2013 Budget Summary and Annual Performance Plan" (2012), accessed October 27, 2012, *http://www.obpa.usda.gov.*

27. *See* Appendix C, table C3.

28. These are (using 2013 USDA budget figures): Farm Service Agency, $12.1 billion; Risk Management Agency, $9.6 billion; Research, Education, and Economics, $2.7 billion; Marketing and Regulatory Programs, $2.4 billion; and Foreign Agricultural Service, $2.1 billion. Note that this budget was prepared in 2012 and is in 2012 dollars, hence these figures require no inflation adjustment. US Department of Agriculture, "FY 2013 Budget."

29. Total 2013 USDA budget for these programs is $28.9 billion. Applying the 63 percent multiplier introduced in chapter 5 as the portion of farm subsidies benefiting animal food production yields $18.2 billion related to animal food production. Physicians Committee for Responsible Medicine, "Agriculture and Health Policies in Conflict."

30. The 2009 study prepared for Canadian dairy farmers estimates 2009 US irrigation subsidies at $1.8 billion (federal) plus $21.5 billion (state) for a total of $23.3 billion, or $24.9 billion in current dollars. (Grey et al., "Farming the Mailbox.") Using the 63 percent multiplier introduced in chapter 5 as the portion of farm subsidies benefiting animal food production, the portion of this total attributable to meat and dairy is $15.7 billion. (Physicians Committee for Responsible Medicine, "Agriculture and Health Policies in Conflict.") The estimated reduction in demand resulting from the Meat

Tax is 65 percent of the tax rate of 50 percent, or 32.5 percent. Multiplying 32.5 percent by \$15.7 billion yields \$5.1 billion. However, land used for livestock or feed crops would likely be repurposed to different crops following a drop in consumption of animal foods, and such subsidies would likely continue—although supporting healthier foods. Accordingly, such a subsidy shift is not counted as a reduction in the subsidy total for purposes of savings calculations.

31. *See* Appendix C, table C4.

32. This figure assumes that a reduction in animal food consumption will cause a pro rata reduction in deaths attributable to eating animal foods. This assumption has clinical support in the research cited in chapter 6, which finds that the less animal foods people eat, the less susceptible to disease they are. Lives-saved figures are based on annual deaths from heart disease (598,607), cancer (568,668), and diabetes (68,504), multiplied by the percentages of these deaths attributed to meat and dairy consumption: 0.3, 0.33, and 0.33, respectively, and further multiplied by the 0.441 decrease in consumption. Death figures are from Kenneth D. Kochanek et al., "Deaths: Preliminary Data for 2009," *US Centers for Disease Control and Prevention National Vital Statistics Reports* 59, no. 4 (2011).

33. Fifty-nine billion land and marine animals are killed for food each year in the United States; 44.1 percent of this total is 26 billion. Free from Harm, "59 Billion Land and Sea Animals Killed for Food in the US in 2009" (2011), accessed August 18, 2012, *http://freefromharm.org.*

34. According to the US EPA, total US carbon dioxide (CO_2) equivalent emissions were 6,821.8 million metric tons (MMT) in 2010. (US Environmental Protection Agency, "U.S. Greenhouse Gas Inventory Report," *Inventory of U.S. Greenhouse Gas Emissions and Sinks: 1990–2010* [2012], accessed October 1, 2012, *http://www.epa.gov.*). Goodland and Anhang estimate that 51 percent of emissions of CO_2 equivalents is attributable to animal agriculture, which represents 3,479.1 MMT of the US total. (Goodland and Anhang, "Livestock and Climate Change.") The 44.1 percent of this figure that the tax proposal would eliminate is 1,534 MMT, or 3.4 trillion pounds of CO_2 equivalents. That is more than the 1,497 MMT that the US EPA estimates was emitted in 2010 by all US passenger cars, trucks, buses, motorcycles, boats, and ships. (US Environmental Protection Agency, "Inventory of U.S. Greenhouse Gas Emissions," 14, table 3.12; note that MMT and teragrams are equivalent units of measure).

35. US animal feed operations generate 500 million tons of manure yearly. Multiply 44.1 percent by that total to get 220 million tons, or 440 billion pounds. US Environmental Protection Agency, "National Pollutant Discharge Regulation Guidelines."

36. The USDA estimates that 35 percent of the total US land area of 2.3 billion acres, or about 805 million acres, is used to graze livestock. (Ruben

N. Lubowski et al., "Major Uses of Land in the United States, 2002," USDA Economic Research Service [2006], accessed August 19, 2012, *http://www.ers.usda.gov.*) Further, while 442 million US acres of nongrazing land is used as cropland, 52 percent of this—or 223 million acres—is dedicated to growing feed crops. (Ibid.) Added to the 805 million acres of grazing land, that's a total of 1,028 million acres. Reducing this total by 44.1 percent to reflect the tax's effect on consumption suggests 453 million acres, or about 708,000 square miles, would no longer be devoted to animal agriculture. And since research finds that it takes five times as much land to feed an omnivore as an herbivore, it should take only one-fifth of that land (about 142,000 square miles) to address the increase in consumption of plant-based foods that will result from the Meat Tax. (Christian J. Peters, Jennifer L. Wilkins, and Gary W. Fick, "Testing a Complete Diet Model for Estimating the Land Resource Requirements of Food Consumption and Agricultural Carrying Capacity: The New York State Example," *Renewable Agriculture and Food Systems* 22, no. 2 [2007]: 145–53.)

37. *See* Appendix C, table C8.

38. *See* Appendix C, table C10.

39. *See* Appendix C, table C9.

40. Khoa Dang Truong and Roland Sturm, "Weight Gain Trends across Sociodemographic Groups in the United States," *American Journal of Public Health* 95, no. 9 (2005): 1602–4.

41. Doreen M. Rabi et al., "Association of Socio-Economic Status with Diabetes Prevalence and Utilization of Diabetes Care Services," BMC Health Services Research, accessed August 14, 2012, *http://www.biomedcentral.com.*

42. Gary L. Francione, *Rain without Thunder: The Ideology of the Animal Rights Movement* (Philadelphia: Temple University Press, 1996); Gary L. Francione and Robert Garner, *The Animal Rights Debate: Abolition or Regulation?* (New York: Columbia University Press, 2010).

43. US Bureau of Labor Statistics, "Highest Incidence Rates of Total Nonfatal Occupational Injury and Illness Cases, 2009," *2009 Survey of Occupational Injuries and Illnesses.*

44. Workers earned $7.70 an hour in 2000; that's $10.24 in 2012 dollars. Charlie LeDuff, "At a Slaughterhouse, Some Things Never Die: Who Kills, Who Cuts, Who Bosses Can Depend on Race," *New York Times* (June 16, 2000).

45. Ibid.

46. Robert Gottleib and Anupama Joshi, *Food Justice* (Cambridge, MA: MIT Press, 2010).

47. Caryl Phillips, *The Atlantic Sound* (New York: Random House, 2000), 33–34.

48. Drew Gilpin Faust, ed., *The Ideology of Slavery: Proslavery Thought in the Antebellum South, 1830–1860* (Baton Rouge: Louisiana University Press, 1981), 21, 30.

49. John Stuart Mill, *On Liberty* (London: Penguin, 1982), 68.

50. Christopher D. Stone, *Should Trees Have Standing? Law, Morality and the Environment* (New York: Oxford University Press, 2010).
51. Cited in Bradford Fitch, "Five Strategies for Building Successful Relationships with Elected Officials," Congressional Management Foundation (2012), accessed October 27, 2012, *www.congressionalfoundation.org*.

Appendix A

1. Beef Checkoff, "Beef Checkoff Hosts Protein Webinars," accessed January 6, 2012, *http://www.beefboard.org*; Pork Checkoff, "Quick Facts: The Pork Industry at a Glance," 3, accessed January 6, 2012, *http://showpig.com*.
2. John McDougall, "Plant Foods Have a Complete Amino Acid Composition," *Circulation* 105 (2002): e197.
3. M. G. Hardinge, H. Crooks, and F. J. Stare, "Nutritional Studies of Vegetarians: V. Proteins and Essential Amino Acids," *Journal of the American Dietetic Association* 48, no. 1 (1966): 25–28.
4. Stanger, *Perfect Formula Diet*, 34.
5. Milton R. Mills, "The Comparative Anatomy of Eating" (2009), accessed October 13, 2011, *http://www.earthsave.ca*.
6. Victor Herbert, "Vitamin B-12: Plant Sources, Requirements and Assay," *American Journal of Clinical Nutrition* 48, no. 3 (1988): 852–58.
7. J. Hultdin et al., "Plasma Folate, Vitamin B12, and Homocysteine and Prostate Cancer Risk: A Prospective Study," *International Journal of Cancer* 113, no. 5 (2005): 819–24.
8. Pat Shipman, "Scavenging or Hunting in Early Hominids: Theoretical Framework and Tests," *American Anthropologist* 88, no. 1 (1986): 27–43.
9. Quoted in Neil Schoenherr, "'Man the Hunter' Theory Debunked in New Book," *Washington University in St. Louis Newsroom* (2006), accessed October 15, 2011, *http://news.wustl.edu*.
10. Donna Hunt and Robert W. Sussman, *Man the Hunted: Primates, Predators, and Human Evolution* (Boulder, CO: Westview Press, 2005).
11. United Nations Food and Agriculture Organization, "Smallholder Dairy Development—Lessons Learned in Asia" (2009), 1, accessed December 29, 2011, *ftp://ftp.fao.org*.
12. Michael Klaper, speech of July 19, 1985, quoted in *Organic Health and Beauty*, "Got Milk," accessed April 1, 2013, *http://www.organichealthandbeauty.com*.
13. Catherine S. Berkey et al., "Dairy Consumption and Female Height Growth: Prospective Cohort Study," *Cancer Epidemiology, Biomarkers & Prevention* 18 (2009): 1881–87.
14. Martin Ahlgren et al., "Growth Patterns and the Risk of Breast Cancer in Women," *New England Journal of Medicine* 351 (2004): 1619–26; B. L. De Stavola et al., "Childhood Growth and Breast Cancer," *American Journal of Epidemiology 159* (2004): 671–82.

15. Campbell and Campbell, *China Study.*

16. *See for example*, E. Giovannucci et al., "Calcium and Fructose Intake in Relation to Risk of Prostate Cancer," *Cancer Research* 58, no. 3 (1998): 442–47; A. G. Schuurman et al., "Animal Products, Calcium and Protein and Prostate Cancer Risk in the Netherlands Cohort Study," *British Journal of Cancer* 80, no. 7 (1999): 1107–13; J. M. Chan et al., "Dairy Products, Calcium, and Prostate Cancer Risk in the Physicians' Health Study," *American Journal of Clinical Nutrition* 74, no. 4 (2001): 549–54.

17. Giovannucci et al., "Calcium Intake in Relation to Prostate Cancer."

18. Jane A. Plant, *The No-Dairy Breast Cancer Prevention Program: How One Scientist's Discovery Helped Her Defeat Her Cancer* (New York: St. Martin's Press, 2001).

19. S. C. Larsson, L. Bergkvist, and A. Wolk, "Milk and Lactose Intakes and Ovarian Cancer Risk in the Swedish Mammography Cohort," *American Journal of Clinical Nutrition* 80, no. 5 (2004): 1353–57; Lawrence H. Kushi et al., "Prospective Study of Diet and Ovarian Cancer," *American Journal of Epidemiology* 49 (1999): 21–31.

20. A. J. Lanou, S. E. Berkow, and N. D. Barnard, "Calcium, Dairy Products, and Bone Health in Children and Young Adults: A Reevaluation of the Evidence," *Pediatrics* 115, no. 3 (2005): 736–43.

21. Ibid., abstract.

22. D. Feskanich et al., "Milk, Dietary Calcium, and Bone Fractures in Women: A 12-Year Prospective Study," *American Journal of Public Health* 87, no. 6 (1997): 992–97.

23. R. G. Cumming and R. J. Klineberg, "Case-Control Study of Risk Factors for Hip Fractures in the Elderly," *American Journal of Epidemiology* 139, no. 5 (1994): 493–503.

24. Uriel S. Barzel and Linda K. Massey, "Excess Dietary Protein Can Adversely Affect Bone," *Journal of Nutrition* 128, no. 6 (1998): 1051–53; J. Lemann Jr., "Relationship Between Urinary Calcium and Net Acid Excretion as Determined by Dietary Protein and Potassium: A Review," *Nephron* 81, supplement 1 (1999): 18–25.

25. Campbell and Campbell, *China Study.*

26. D. Mark Hegsted, "Fractures, Calcium, and the Modern Diet," *American Journal of Clinical Nutrition* 74, no. 5 (2001): 571–73.

27. Ibid., 571.

28. Mark Hegsted, "Calcium and Osteoporosis," *Journal of Nutrition* 116, no. 11 (1986): 2316–19.

29. B. J. Abelow, T. R. Holford, and K. L. Insogna. "Cross-Cultural Association between Dietary Animal Protein and Hip Fracture: A Hypothesis," *Calcified Tissue International* 50, no. 1 (1992): 14–18.

30. Martin Ahlgren et al., "Growth Patterns and the Risk of Breast Cancer in Women," *New England Journal of Medicine* 351 (2004): 1619–26.

31. Stavola et al., "Childhood Growth and Breast Cancer."
32. Benjamin Spock, *Dr. Spock's Baby and Child Care*, seventh ed. (Pocket Books, New York: 1998), 195.
33. Quoted in "Does It Taste Good? Watch Out!" *Deseret News* (September 30, 1992), accessed July 14, 2012, *http://www.deseretnews.com*.
34. National Academy of Sciences, "Dietary Reference Intakes for Fat," 441, 542.
35. *See, for example*, R. M. Weggemans, P. L. Zock, and M. B. Katan, "Dietary Cholesterol from Eggs Increases the Ratio of Total Cholesterol to High-Density Lipoprotein Cholesterol in Humans: A Meta-Analysis," *American Journal of Clinical Nutrition* 73, no. 5 (2001): 885–91.
36. Adnan I. Qureshi et al., "Regular Egg Consumption Does Not Increase the Risk of Stroke and Cardiovascular Diseases," *Medical Science Monitor* 13, no. 1 (2007): CR1–8.
37. USDA Agricultural Research Service, "Nutrient Intakes from Food: Mean Amounts Consumed per Individual, One Day, 2005–2006 (2008)," accessed January 26, 2012, *http://www.ars.usda.gov*.
38. Norris and Messina, "Disease Markers of Vegetarians."
39. Yasuyuki Nakamura et al., "Egg Consumption, Serum Cholesterol, and Cause-Specific and All Cause Mortality: The National Integrated Project for Prospective Observation of Non-Communicable Disease and Its Trends in the Aged, 1980 (NIPPON DATA80)," *American Journal of Clinical Nutrition* 80 (2004): 58–63.
40. Ibid., 63.
41. Poultry Production News, "American Egg Board Funded $2 Million in Nutrition Research in 2010" (2011), accessed December 31, 2011, *http://poultry-productionnews.blogspot.com*.
42. Marcia D. Greenblum, "An Egg a Day is More than Okay!" Nutrition Realities, accessed December 31, 2011, *http://www.eggnutritioncenter.org*.
43. Penny M. Kris-Etherton, William S. Harris, and Lawrence J. Appel, "Fish Consumption, Fish Oil, Omega-3 Fatty Acids, and Cardiovascular Disease," *Circulation* 106 (2002): 2747–57.
44. B. C. Scudder et al., "Mercury in Fish, Bed Sediment, and Water from Streams across the United States, 1998–2005," US Geological Survey Scientific Investigations Report 2009–5109 (2009).
45. David L. Stalling and Foster Lee Mayer Jr., "Toxicities of PCBs to Fish and Environmental Residues," *Environmental Health Perspectives* 1 (1972): 159–64.
46. Campbell and Campbell, *China Study*.
47. Scudder et al., "Mercury in Fish."
48. E. Sunderland et al., "Mercury Sources, Distribution, and Bioavailability in the North Pacific Ocean: Insights from Data and Models," *Global Biogeochemical Cycles* 23 (2009).
49. Stalling and Mayer, "Toxicities of PCBs to Fish."

50. Consumeraffairs.com, "Americans Confused about Health Effects of Eating Fish" (2006), accessed November 15, 2011, *http://www.consumeraffairs.com*; Charlotte Seidman, "Fish is Good—Fish is Bad. Balancing Health Risks and Benefits," *American Journal of Preventive Medicine* (2005), accessed November 15, 2011, *http://foodconsumer.org*.

51. Janet M. Torpy, Cassio Lynm, and Richard M. Glass, "Eating Fish: Health Benefits and Risks," *Journal of the American Medical Association* 296, no. 15 (2006): 1926.

52. Environmental Defense Fund, "List of Seafood Health Alerts," accessed November 15, 2011, *http://apps.edf.org*.

Appendix B

1. Health care, see chapter 6; subsidies, see chapter 5; environmental, see chapter 7; cruelty, see chapter 8; fishing, see chapter 9; inflation adjustments, US Department of Labor Bureau of Labor Statistics, "CPI Inflation Calculator," accessed October 27, 2012, *http://www.bls.gov*.

Appendix C

1. US Poultry & Egg Association, "Economic Data" (2011), accessed August 13, 2012, *http://uspoultry.org*; USDA National Agricultural Statistics Service, "Milk Production, Disposition and Income 2011 Summary" (2012), accessed August 13, 2012, *http://usda.mannlib.cornell.edu*; USDA National Agricultural Statistics Service, "Meat Animals Production, Disposition and Income 2011 Summary" (2012), accessed August 13, 2012, *http://usda.mannlib.cornell.edu*.

2. Tatiana Andreyeva, Michael W. Long, and Kelly D. Brownell, "The Impact of Food Prices on Consumption: A Systematic Review of Research on the Price Elasticity of Demand for Food," *American Journal of Public Health* 100, no. 2 (2010): 216–22.

3. These are (using 2013 USDA budget figures): Farm Service Agency, $12.1 billion; Risk Management Agency, $9.6 billion; Research, Education, and Economics, $2.7 billion; Marketing and Regulatory Programs, $2.4 billion; and Foreign Agricultural Service $2.1 billion.

4. 63 percent of subsidies are related to animal food production ($28.9 billion x 0.63 = $18.2 billion). Physicians Committee for Responsible Medicine, "Agriculture and Health Policies in Conflict."

5. One view is that American farmers pass through to consumers "less than ten percent" of corn-related cost increases (Ephraim Leibtag, "Corn Prices Near Record High, But What about Food Costs?" *Amber Waves: The Economics of Food, Farming, Natural Resources and Rural America* [2008]). Another is that American consumers pay "the bulk of" cost increases related to livestock production (Bruce Gardner, "The Economic System of

U.S. Animal Agriculture and the Incidence of Cost Increases," in *Sharing Costs of Change in Food Animal Production: Producers, Consumers, Society and the Environment*, ed. Richard Reynnells [Washington, DC: USDA, 2003] 6).

6. $115.1 billion x 0.32 = $36.8 billion. US Bureau of Labor Statistics, "Consumer Expenditure Survey 2010" (2011), Current Expenditure Tables, accessed December 3, 2011, *http://www.bls.gov*.

7. As explained in chapter 10, because we've seen that checkoffs drive about $4.6 billion in annual sales of animal foods, or 1.8 percent of the industry's total annual sales of $251 billion, eliminating these programs should reduce consumption by about 1.8 percent.

8. Figures adjusted for inflation. US Bureau of Labor Statistics, "Consumer Expenditure Survey 2010."

9. Joint and separate married filers combined as married couple households. Internal Revenue Service, "SOI Tax Stats—Individual Statistical Tables by Filing Status" (2009), accessed August 12, 2012, *http://www.irs.gov*.

10. Centers for Medicare and Medicaid Services, Office of the Actuary, "National Health Expenditure Projections 2010–2020," table 3, accessed August 13, 2012, *http://www.cms.gov*.

Appendix D

1. Charles Dickens, *Dombey and Son* (Hertfordshire, UK: Wordsworth Editions, 1995).

2. Jeffrey Moussaieff Masson, *The Pig Who Sang to the Moon: The Emotional World of Farm Animals* (New York: Ballantine, 2003), 27.

3. Scully, *Dominion*.

4. Ibid.

5. Ibid., 267–68.

6. Quoted in Joanne Stepaniak and Virginia Messina, *The Vegan Sourcebook* (Los Angeles: Lowell House, 1998), 39.

7. R. Nowak, *Walker's Mammals of the World 5.1.* (Baltimore: Johns Hopkins University Press, 1997); Richard L. Wallace, "Market Cows: A Potential Profit Center," University of Illinois Extension (2002), accessed August 6, 2011, *http://www.livestocktrail.uiuc.edu*.

8. Cows have a nine-month gestation cycle, produce milk for about ten months after giving birth, and are re-impregnated about two months into the cycle. This adds up to about one pregnancy—and one newborn calf—per year for each cow.

9. Quoted in Mercy for Animals, "Vegetarian Starter Kit," 11, accessed September 28, 2012, *http://www.mercyforanimals.org*.

10. Jason Henderson and Ken Foster, "Characteristics of U.S. Veal Consumers" (staff paper, 00-2, Department of Agricultural Economics, Purdue University, 2000), accessed September 28, 2012, *http://ageconsearch.umn.edu*.

11. James M. MacDonald et al., "Profits, Costs, and the Changing Structure of Dairy Farming," *Economic Research Report* ERR-47 (September 2007).
12. A. L. Legrand, M. A. G. von Keyserlingk, and D. M. Weary, "Preference and Usage of Pasture Versus Free-Stall Housing by Lactating Dairy Cattle," *Journal of Dairy Science* 92, no. 8 (2009): 3651–58.
13. Quoted in Animal Aid, "Battery Cows: Zero Grazing and the Dairy Industry," accessed August 11, 2011, *http://www.animalaid.org.*
14. Quoted in Jonathan Leake, "The Secret Life of Cows" (2005), accessed July 14, 2012, *http://www.rense.com.*
15. Discussed in Ibid.
16. Quoted in Compassion over Killing, "A COK Report: Animal Suffering in the Broiler Industry," accessed August 5, 2011, *http://www.cok.net.*
17. D. Martin, "Researcher Studying Growth-Induced Diseases in Broilers," *Feedstuffs* (May 26, 1997).
18. S. C. Kestin et al., "Prevalence of Leg Weakness in Broiler Chickens and Its Relationship with Genotype," *Veterinary Record* 131, no. 9 (1992): 190–94.
19. T. C. Danbury et al., "Self Selection of the Analgesic Drug Carprofen by Lame Broiler Chickens," *Veterinary Record* 146, no. 11 (2000): 307–11.
20. Quoted in J. Erlichman, "The Meat Factory: Cruel Cost of Cheap Pork and Poultry—Factory Methods Have Slashed Meat Prices in the Last 30 Years," *The Guardian* (October 14, 1991).
21. C. A. Weeks et al., "The Behaviour of Broiler Chickens and Its Modification by Lameness," *Applied Animal Behaviour Science* 67, no. 1 (2000): 111–25.
22. Inma Estevez, "Poultry Welfare Issues," *Poultry Digest Online* 3, no. 2 (2002), accessed September 28, 2012, *http://ansc-test.umd.edu.*
23. Marcus, *Meat Market*, 22.
24. Egg producers added nearly 270 million laying hens to their flocks in 2010. Assuming that an equal number of males and females are born, that's roughly the number of male chicks killed. USDA National Agricultural Statistics Service, "Chicken and Eggs 2010 Summary" (February 2011).
25. Michael J. Gentle et al., "Behavioural Evidence for Persistent Pain Following Partial Beak Amputation in Chickens," *Applied Animal Behaviour Science* 27, no. 1 (1990): 149–57.
26. Ibid.
27. Peter Singer and James Mason, *Animal Factories* (New York: Crown, 1980).
28. *Collins English Dictionary*, online edition, accessed September 28, 2012, *http://www.collinsdictionary.com; FindLaw Legal Dictionary*, online edition, accessed September 28, 2012, *http://dictionary.findlaw.com.*
29. Valerie Brewer, "An Introduction to Chicken Production: A Brief Insight into the Modern Chicken and Egg Industries," National Chicken Council (2007), accessed September 28, 2012, *http://www.ca.uky.edu.*

30. Milton H. Arndt, *Battery Brooding: A Complete Exposition of the Important Facts Concerning the Successful Operation and Handling of the Various Types of Battery Brooders* (Chicago: Orange Judd Publishing Company, Inc., 1931).

31. Ibid.

32. Marcus, *Meat Market*.

33. US Poultry and Egg Association, "Economic Data" (2010), accessed July 26, 2011, *http://www.poultryegg.org*; USDA National Agricultural Statistics Service, "Chicken and Eggs 2010 Summary," February 2011.

34. Ian J. H. Duncan, "Animal Welfare Issues in the Poultry Industry: Is there a Lesson to Be Learned?" *Journal of Applied Animal Welfare Science* 4, no. 3 (2001): 207–21.

35. United Poultry Concerns Inc., "The Animal Welfare and Food Safety Issues Associated with the Forced Molting of Laying Birds," accessed September 19, 2012, *http://www.upc-online.org*.

36. American Egg Board, "Factors that Influence Egg Production" (2010), accessed September 28, 2012, *http://www.aeb.org*.

37. Jeffrey Moussaieff Masson, *The Face on Your Plate: The Truth About Food* (New York: W. W. Norton & Company, 2009).

38. Lesley J. Rogers, *The Development of Brain and Behaviour in the Chicken* (Wallingford, UK: CAB International, 1995).

39. Konrad Lorenz, "Animals are Sentient Beings: Konrad Lorenz on Instinct and Modern Factory Farming," *Der Spiegel* (November 17, 1980).

40. Marian Stamp Dawkins, *Through Our Eyes Only?—The Search for Animal Consciousness* (New York: Oxford University Press, 1998).

41. Masson, *The Face on Your Plate*.

42. Ibid.

43. American Egg Board, "Facts about the Egg Production Process" (2010), accessed July 28, 2011, *http://www.aeb.org*; Elizabeth Weise, "Cage-Free Hens Pushed to Rule Roost," *USA Today* (April 10, 2006).

44. Born Free, "Progressive Farming," accessed July 30, 2011, *http://www.bornfreeeggs.com*.

45. Pollan, *Omnivore's Dilemma*.

46. Jewel Johnson, "A Rare Glimpse Inside a Free-Range Egg Facility," *Prairie Progress* 8, no. 8 (2007), accessed July 31, 2011, *http://www.peacefulprairie.org*.

47. Masson, *The Face on Your Plate*.

48. Jennifer Welsh, "Hens Feel for Their Chicks' Discomfort," *LiveScience* (March 9, 2011).

49. Rogers, *Brain and Behaviour in the Chicken*.

50. W. Grimes, "If Chickens Are So Smart, Why Aren't They Eating Us?" *New York Times* (January 12, 2003).

51. Bernard E. Rollin, *Farm Animal Welfare: School, Bioethical, and Research Issues* (Ames, IA: Iowa State University Press, 1995), 118.

52. Karen Davis, "My Experience of Empathy and Affection in Chickens and the Social Life of Chickens and the Mental States I Believe They Have and Need in Order to Participate in the Social Relationships I Have Observed," United Poultry Concerns (2009), accessed August 7, 2011, *www.upc-online.org*.

53. California Health and Safety Code §§ 25990-25994.

54. Humane Society of the United States, "Today, The HSUS and the United Egg Producers Announced an Agreement which Could Result in a Complete Makeover of the U.S. Egg Industry," Humane Society Press Release (August 6, 2011), accessed August 6, 2011, *http://action.humanesociety.org*.

55. Nedim C. Buyukmihci, "A Veterinarian's Perspective on the Rotten Egg Bill" (2012), accessed July 20, 2012, *http://stoptherotteneggbill.org*.

INDEX

Fitzgerald, Peter, 65, 174
Food and Drug Administration
Bovine growth hormone (rBST), 59–60
Regulation of animal antibiotics, 60–62
Scope of responsibility for animal foods, 56
Food miles, 121–123
Forced molting, 221–222
Francione, Gary, 178
Freedman, Rory, 169
Frost, Robert, xxi

G

Galbraith, John Kenneth, xiii
Gandhi, Mohandas, 183, 185
Garner, Joseph, 137–138
German, Mike, 44–45
Godwin, Jerry, 214
Goldfish, capacity to feel fear and anxiety, 137–138
Goodland, Robert, 125
Gottleib, Robert, 180
Gout, 94
Government speech, 6
Grandin, Temple, 47
Green Scare, 42–45
Ground beef
Cholesterol content, 93
E. coli, 103

H

HAACP. See Hazard Analysis Critical Control Points
Halal, 47
Ham, 26, 96
Historical price movement, 74
Harkin, Tom, 129
Hayes, Rutherford B., 164
Hazard Analysis Critical Control Points, 65
Heart disease
Annual US health care cost, 107
Chicken consumption, 92
Incidence in US, 90
Incidence related to animal food production, 96
Red meat consumption, 20, 93

Hegsted, D. Mark, 195–196
Hens
Battery cages, 220–221
Cage-free conditions, 223–225
Campylobacter, 102–103
Consumer willingness to pay to move to free-range system, 141
Enriched cages, 227–228
Forced molting, 221–222
Free-range, 223–225
Improvements in production, 74
Inability to roost in factory farms, 225–227
Intelligence and personality, 225–227
Organic production, 115–118
Partial beak amputation, 218–226
Salmonella, 103
Slaughter methods, 37
Treatment in factory farms, generally, 218–228
HMSA. See Humane Methods of Slaughter Act
Hogs. See Pigs
Hormones in beef and dairy, effects on human health, 106
Hughes, Arthur, 49
Humane Methods of Slaughter Act, 24, 47–50, 52
Humane Society of the United States, 44, 49, 227
Huxley, Aldous, 3, 14

I

IMTA. See Integrated Multi-Trophic Aquaculture
Incidental taking. See Bycatch
Integrated Multi-Trophic Aquaculture, 150–151
Iron, 191

J

Jefferson, Thomas, 53
Jevons Paradox, 155
Johnson, Jewel, 224
Johnson, Stephen L., 128
Joshi, Anupama, 180
Joy, Melanie, xxiii

Supply-driven forces, 96, 100, 116, 145, 155, 159, 164
Sussman, Robert, 192
Sustainability of animal food production, 117–121
    Air pollution, 112, 126
    Climate change, 125
    Ecological rotation, 114–117
    Fertilizers, 125–126
    Generally, 131–132
    Local production, 121–123
    Organic production, 117–121
    Pesticides, 125–126
    Water pollution, 126–127
    Water use, 118–120
Swine. See Pigs
Swine flu, 16–18

T

Tax on animal foods. See Meat Tax, proposed
Tax credit, proposed, 173
Terrorism, 43–44
Tilapia, 154–156
Tobacco industry, xx-xxi, 109, 177
Torell, Ron, 135
Trout
    Capacity to feel pain, 136–137
    Slaughter methods, 139–140
    Stocking density in fish farms, 138–139
    Stress related to tight stocking density, 138–139
    Unsustainability of farming operations in Sweden's coastal waters, 156
Tuttle, Will, xix
Twain, Mark, 17, 173
Tyson Foods, 19, 48, 87

U

US Department of Agriculture
    Beef recall, 49–50
    Cholesterol guidelines, 63, 79
    Enforcement of Humane Methods of Slaughter Act, deficiencies in executing, 48–50
    Food Safety and Inspection Service, 52

Inability to issue mandatory recall, 66
Inherent conflicts of interest, 62–69
Inspection duties, deficiencies in executing, 48–50
Involvement in checkoff programs, 6
Labeling duties, deficiencies in executing, 66–68
Office of Inspector General, 65, 66
Protein guidelines, 26, 29–30
Reduction in subsidy payments, proposed, 175
Reform, proposed, 174
Reminded by Congress to enforce Humane Methods of Slaughter Act, 49
Revolving door of industry personnel, 68–69
Saturated fat guidelines, 101–102
Scope of responsibility for animal foods, 56
Secretary Tom Vilsack, 17, 68–69
United Poultry Concerns, 222, 226
Unruh, Jesse "Big Daddy", 33
USDA. See US Department of Agriculture

V

Vanderbilt, William, 52
Veal calves, treatment in factory farms, 216
Vegetables, quality of protein, 198–200
Vilsack, Tom, 17, 68–69
Vitamin B_{12}, 191–192
Voit, Carl von, 29

W

Warhol, Andy, 166
Water pollution caused by animal food production, 126–127
Wilde, Oscar, 114
Wilson, Woodrow, 163
Winfrey, Oprah, xvii, 38
Wolfson, David, 32, 36
Wordsworth, William, 184
Wynn, Steve, 109

ABOUT THE AUTHOR

David Robinson Simon is a lawyer and advocate for sustainable consumption. He works as general counsel for a healthcare company and serves on the board of the APRL Fund, a non profit dedicated to protecting animals.

David received his BA from UC Berkeley and his JD from the University of Southern California. He is also the author of *New Millennium Law Dictionary*, a full-length legal dictionary. He lives in Southern California with his partner, artist Tania Marie, and their rabbit, tortoise, and two cats.

Visit David at *www.meatonomics.com*.